CON NEX IONS

THE UNSEEN HAND OF TONY HUNT

NIGEL DALE

Whittles Publishing

Published by
Whittles Publishing,
Dunbeath,
Caithness KW6 6EG,
Scotland, UK
www.whittlespublishing.com

© 2012 Nigel Dale
ISBN 978-184995-030-5

Stencil Gothic BE ©1999 by Brad O. Nelson/Brain Eaters Font Co. & Jeff Levine.

Printed by Finidr Ltd., Czech republic

CONTENTS

The publishers would like to thank the following
for their sponsorship of this volume

Sinclair Knight Merz (SKM) is delighted to be associated with Tony Hunt's biography:
Connexions - the Unseen Hand of Tony Hunt.

A leading projects firm, with global capability in strategic consulting, engineering and project
delivery, SKM has had a long and trusted relationship with Tony Hunt for many years and it is a
great privilege for us to be associated alongside some of his greatest achievements.

In 2004, SKM merged with Anthony Hunt Associates, one of Britain's leading structural and
civil engineering consultancies. SKM and Anthony Hunt Associates shared a common design
ethos towards elegant and economical building and civil engineering structures based on a
strong philosophy of early collaboration across the design team. SKM would like to congratulate
Tony on his many project achievements. We are incredibly proud to have worked so closely
with him over the years and to have been associated with a number of the well recognised
buildings of our time that are credited in this book. These developments include the world
famous Eden Project in Cornwall, The Roundhouse in Camden and The New Area Terminal at
Barajas Airport in Madrid

SKM is an organisation with a proud history where we embrace our shared values and an
open culture. We have a commitment to service and quality, with high standards of safety and
business ethics, along with a leading edge approach to delivering a sustainable future.

We operate in three regions: Asia Pacific, the Americas and EMEA (Europe, Middle East &
Africa), deploying some 6,500 people from more than 40 offices while serving the Buildings
and Infrastructure, Mining and Metals, Power and Energy and Water and Environment sectors.
Formed in 1964 in Sydney as a private company, SKM has retained its independence through
employee ownership, with fee income now greater than £700 million.

Richard + Ruth Rogers

JOHN McASLAN + PARTNERS

AUTHOR'S NOTE

I graduated from Manchester University School of Architecture in the 1970s. There, great emphasis was placed upon the History of Architecture. Professor RA Cordingley (who at that time had recently revised *A History of Architecture on the Comparative Method* – a standard work written by Banister Fletcher) had recently left the School. Donald Buttress and Ronald Brunskill (experts in Gothic and Gothic Revival architecture and English timber-framed buildings respectively) lectured in construction, conveying to us the importance of architectural history. But in the wider sphere, there were visiting lecturers: Paolo Soleri (who spoke about his visions of alternative communities in space) and Peter Cook of Archigram (who talked about 'Instant Cities') were both memorable, as was an exhibition of Buckminster Fuller's proposal for a huge geodesic dome to cover Manhattan.

Around the School of Architecture, the informal talk was of a student from several years earlier: one Norman Foster who had left Manchester for a successful architect's life in London. His offices were the dream as far as a 'year out' was concerned. At Manchester any over-exuberance or outlandish dress, or especially any unheralded offbeat designs were greeted with a quiet word: 'Do you think you might be more suited to the Architectural Association'? I remember being told that a design I had developed for a glass house that tracked the position of the sun by mechanical means, was no more than 'a fun object' and that I should tear up that design and start again.

Word had first reached me about Michael and Patricia Hopkins from my late father. He was acquainted with Dennis Wainwright – a renowned orthopaedic surgeon living nearby in Staffordshire. Wainwright told my father that his daughter Patricia was studying at the Architectural Association in London, and that the working environment was both challenging and rewarding. On one occasion during the Patera Building time, I took Patti to visit the much-changed house in Clayton where she grew up. Dennis Wainwright had sold most of his garden in order for the new Nuffield hospital to be built. What was left had been converted into offices.

I first met Tony Hunt in early 1980. It was in the studio of Michael Hopkins' house at Downshire Hill, Hampstead. I had approached Hopkins about forming a commercial partnership to develop the SSSALU (Short Span Structures in Aluminium) aluminium building system that Patti and Michael Hopkins, together with Ian Ritchie and Tony Hunt, had designed. We met on the occasion of a second meeting at the

Hopkins' House to discuss the matter, and it was there that Tony Hunt was introduced to me as 'a structural engineer specialising in lightweight structures'. Discussions moved quickly beyond the SSSALU system, and the all-steel Patera Building was born. During the next two years, there were many meetings with Tony Hunt both in London and in Stoke-on-Trent where the Patera Building was prototyped and launched. During this time, Tony Hunt was going through his second divorce from Pat Hunt, he was selling the much-cherished Coln Manor – the Anthony Hunt Associates (AHA) offices in Gloucestershire which doubled as his home – and he was dating Diana Collett, who was to become his third wife.

Whilst working on the design for the Patera Building, conversations ranged from furniture design to car design, from science-fiction to childhood mechanical toys, from steel processing to the latest welding and weld testing techniques, and all sorts of related subjects in between. Even then, the influences of Richard Buckminster Fuller, the Californian school building systems, the work of West Coast American architects and Charles and Ray Eames were seen as influences upon Hunt's work. AHA produced a superb set of structural design drawings, structural calculations and component drawings for the Patera Building. Mark Whitby and Alan Jones of AHA were instrumental in some of the more detailed work. At Patera we were given an insight into the world of structural engineering, with a glimpse into the design problems associated with the British Antarctic Survey Halley Bay designs. In hindsight, with all the uncertainties of Tony Hunt's personal life, it can be seen as testament to his relaxed authority and unrivalled experience in lightweight steel design that any upheaval in his personal life went quite unnoticed.

In 2007 I was asked to write an abridged biography of Tony Hunt and during one of the meetings to discuss the scope of that writing, Tony suggested that I should write a fuller version. During a series of daylong discussions over the last three years I have been able to put together a story of the events, the meetings, the people, and of course, details of projects that Tony has worked on throughout his career. I have had a book title in mind for many years but under the general scope of the 'History and Origins of the High-Tech Movement in Britain'. Tony Hunt's story impacts upon the general 'History' in almost every aspect, and one cannot write about the High-Tech Movement without constant reference to Hunt. So in a way, this book has enabled me to combine the two themes.

Nigel Dale

FOREWORD BY SIR JAMES DYSON

I was about as ignorant as you can be about design when I arrived at the Royal College of Art in 1966, and despite the name, it is in fact 80% design. Fortunately I was able to spend an introductory year visiting most of the design and art departments. It was architecture that really caught my fancy. I liked the scale. But architecture involves structural engineering, and I had never met an engineer. Who were these backroom boys?

All was explained when I came across Tony Hunt. Here was somebody who didn't see structural necessity and creative design as mutually exclusive. It was quite a revelation. I had spent my formative years with the assertion thrust upon me again and again that engineering and art were two very different concepts never to be combined.

This bludgeoned approach never sat very comfortably with me, but nonetheless it was the message at every corner. So when I came across Tony talking with equal enthusiasm about design and the theory of structures, I was taken aback. This was an art far removed from that taught in the classroom. Art it was though, as he drew his structures with such a sure hand.

He championed the likes of Buckminster Fuller as 'dreamers'. Dreamers, I'd been taught, were lazy, ideological romantics and never got anything done. The reality, Tony showed us, is that dreamers were visionaries that could conceive a world that did not yet exist.

That it has taken quite so long for such a comprehensive study of Tony's work to surface is telling. His ambition and desire to experiment are without parallel in the last 40 years of structural engineering, but too often this is overlooked. While an architect takes the glory and the headlines, the structural engineer is often assumed to only be doing their job if the place is still standing a hundred years down the line.

Tony has repeatedly shown that form and function are always intertwined. It is the way that something is put together that really enchants the onlooker. Brunel knew that – fashion and fad aside – his mathematically beautiful design would capture the imagination for years to come. That sentiment really gripped me when Tony introduced it to me at the RCA. During a lifetime at different institutions, he has continued to introduce it to fortunate students ever since.

When Dyson outgrew its site in the mid 1990s, I asked Tony and Chris Wilkinson to design us a new one. When you plan to put something quite so grand in the heart of the Cotswolds the temptation is to play it safe. Thankfully, Tony and Chris had other ideas and came up with a building that was both imaginative and modern. It remains to this day a wonderful exhibition of how the structure can lead the design. It becomes the aesthetic. At the heart of all of Tony's work is that very basic assertion.

Within the pages of this book you'll find yourself enthralled by the effectiveness and originality of his work. Time and again he has brought structures of astounding ingenuity to life. Structural engineers that dream on the scale of Tony Hunt are few and far between. Having said that, he's done a very good job of inspiring a few more.

INTRODUCTION

Tony Hunt is unique amongst structural engineers in his passion for industrial design. In his mind's eye, structures can be made up from a series of batch-produced components akin to (but on a larger scale than) furniture manufacture. This industrialised approach to structural design was well received by architects up to the mid 1980s, when system-built prefabricated buildings were in vogue. His vision re-emerged in a number of prestigious projects later in his career when the demand from architects was for the spectacular rather than for the functional or utilitarian.

Hunt's career is the story of six decades of British architecture from the 1950s to the present. He has worked with nearly all the acclaimed British architects during that time. Direct comparisons with the great European and American architect engineers such as Richard Buckminster Fuller, Charles Eames, Fritz Haller and Jean Prouvé are justifiably made. If Sir Ove Arup, Sir Owen Williams and Felix Samuely were the first in a line of engineers working in the UK who in their various ways adopted a bi-disciplinary role of collaboration with British architects, then the following generation of similarly minded engineers would include Frank Newby, Peter Rice, Ted Happold and Tony Hunt. Hunt is the only one of these who is still around at the time of writing.

Hunt's archive is based purely on pictures: photographs, graphic works, sketches, computer-generated images, models and artworks. He has never kept written material. Details of meetings, minutes of committee work, written briefs, letters, academic papers etc have not travelled with him. Hunt has tens of thousands of images, and he has a personal library of many hundreds of books and publications covering the history of art, architecture, biography, furniture and product design. Not all of these images and photographs are dated, so it has been something of a challenge to set in chronological order the sequence of events, meetings and projects.

Anthony Hunt Associates (AHA) – the practice that carried Tony Hunt's name – had a prolific output, particularly in the 1960s and 1970s. Projects were numbered chronologically as they came into the office, and by 1983 the 1,000th job number had been allocated.

The profile of the engineer has diminished due to the rise of the internationally-renowned combined architect and engineering practices, and it is hard to see whom will succeed Hunt as the engineer consistently in tune with the aspirations of British architects and who was uniquely placed to exert such a strong influence over them for a 50 year period.

1   DISRUPTED EARLY YEARS

Some of Tony Hunt's earliest recollections are of wartime incendiary bombs and fires in the Dockland areas of London. He was born in Streatham Hill, London on June 22 1932. His father was a solicitor's clerk and his mother the daughter of teachers. Aged seven when WWII started and with a brother just one month old, the Hunt family moved out of London.

Portrait of Tony as a youngster whilst living in London.

Tony Hunt tells us that he holds very few memories of the first six years of his life. What he does remember sticks out vividly. He remembers visits to a store called Pratts in Streatham, south London. It was a shop like others of the time which seemed to sell everything. Hunt refers to his fascination with a transit system there that linked each sales counter with a central glazed booth at a higher level. Money was put into a capsule at the sales desk and was fired up to this booth travelling along the overhead wires. The capsule was duly returned to the sales assistant with any change and a receipt. The mechanics of the system fascinated the youngster.

Hunt also remembers his grandfather teaching him to ride on a bicycle he had bought for his grandson. These lessons took place on Tooting Bec Common. The memories are of that exhilarating mixture of excitement and terror and of wobbles before mastering his balance. On one occasion Hunt was tearing down a local hill on the pavement when he ran into an old lady, hospitalising her and resulting in him being 'grounded' for one week. Other early memories are of visits to a tea shop run by two Scottish spinster ladies, to another tea shop called Zeeta's, to a marvellous sweet shop called Boynes – where all the sweets were contained in tall jars in the window – and of course, visits to Woolworths for small gifts.

With the start of WWII everything changed and Tony Hunt's early pattern of life was also to change. He remembers the first big docks bombing raid, 'where the whole of East London seemed to be on fire'. Soon afterwards Tony's father decided that the family should move out of London. They went briefly to New Malden – Hunt does not really know why they went there. He attended a school there for a brief time, which

[Left] *Portrait of Tony's Father –
a solicitor's clerk.*

[Below] *Tony's brother Simon
Hunt, seven years his junior.*

he found completely alien. Up to the time of the move out of London he had started his education at a Catholic convent, as both his parents were Catholics.

From New Malden, Tony's brother, his mother and he moved to Burnham-on-Sea, leaving his father in London. From there they moved for a brief time to Hazelmere but after that they settled in Farnborough, Hampshire. A quest for stability in Tony's education may have been the reason for this move, as there was a suitable school there – the Salesian College. Tony remained there until the age of 16, firstly as a dayboy, then with one year as a weekly boarder and finally he spent the last year there as a full time boarder. This became necessary as WWII had ended, allowing his brother and parents to move back to London. Tony's brother Simon became a flautist and Simon's wife a cellist.

Tony Hunt attended the Salesian College, remaining there until he reached the age of 16. Tony Hunt reflects upon his school days as being disrupted and somewhat lonely, but during his time in Farnborough he started making and flying model aircraft. He recalls a friend named Tom Crowdy, with whom he shared this interest. Tom's father worked on a secret project at the Royal Aircraft Establishment. With this association, models of aircraft allowed just a short stretch of the imagination to become reality: this was to become closely tied into the design methods he would use later in his professional life.

The school offered no arts teaching and had no science sixth form, causing Hunt to move again via Northampton Polytechnic – where he studied for less than the academic year – to eventually attend Westminster Technical College for a day release course in Civil Engineering. By the age of 16, Tony Hunt had attended seven different schools. Hunt successfully gained his Civil Engineering qualifications at Westminster Tech.

At this time Hunt's father arranged for the young Tony to be articled through the Worshipful Company of Founders to the small civil engineering City firm of Wheeler and Jupp. As an articled pupil, he was locked in for a period of four years. Hunt describes the firm's work as 'run of the mill', and feels that he didn't learn very much during his time there.

However, at the end of the articled period Hunt moved again to work for a firm of water engineers. He stayed with them only for six months, but whilst working there he had the good fortune to attend a week-long residential course held by the Cement and Concrete Association at Wexham Springs, Slough in Buckinghamshire. Tony Hunt found himself in a stimulating environment here, where interesting structures and technology were discussed, information shared, and designs evaluated.

2 1951, FESTIVAL OF BRITAIN AND WORKING FOR SAMUELY

In 1951 the Festival of Britain – built on the South Bank in London – was an inspired event intended to provide much needed relief from the austerity of the WWII years. It was a political gesture to restore pride in the country at large, when factory production established during the war years was adjusted for peaceable activities and the population returned to normal jobs and pastimes after several years of focus on the war effort. Sir Hugh Casson (1910–1999) was appointed as the Director of Architecture for the Festival. He brought together a group of young energetic architects and engineers who were able to engender a spirit of optimism and futuristic imagery into the Festival buildings. Although the Festival only lasted for a matter of months (nearly all the buildings were torn down on the instruction of the incoming government of 1952), their designs had an impact that stimulated and influenced architects and engineers for future generations.

Amongst the various buildings designed for the Festival was the Dome of Discovery – designed by a then young architect Ralph Tubbs who had recently graduated from the Architectural Association School of Architecture in London, and engineered by the experienced founding partner of Freeman Fox and Partners, Ralph Freeman. With a long escalator providing the dramatic form of entry, the Dome of Discovery was a gleaming aluminium circular structure 111m in diameter, standing 27 metres high. As a focal point for the Festival of Britain in 1951, the building contained a range of themed exhibitions that 'would tell a story about what Britain had done in the past'.

Ralph Freeman – a founding partner of Freeman Fox and Partners (the consultant engineers responsible for the design of the Sydney Harbour bridge) – joined forces with the architect Ralph Tubbs. This association of designers brought mixtures of youth and experience (Freeman was 70 years old and Tubbs 36), and of innovation (Tubbs had graduated from the Architectural Association), coupled with Freeman's conservative engineering expertise. The result was a novel structure, which was speedily erected and which was supplied at an economical price. As a one-off system building, Tubbs described the design process: 'the dome was a kind of mathematical poem'. The completed building presented images of a future science-fiction world – it was depicted in *The Eagle* (a sci-fi comic of the time) and provided relief from the austerity if the post WWII era and an optimistic outlook for the future. The comments of Ralph

Tubbs that 'the dome was a kind of mathematical poem' was an echo heard down the years from comments made by the acclaimed novelist Neville Schute Norway. He has described how after months of intense work in the calculation (using seven figure log tables) of tensile structures within the frame of the R100 airship, he also experienced some sort of spiritually uplifting feeling.

Considerable technological innovation was required in the production and shaping of aluminium extrusions to provide components for the structure. Progressions of triangular bays were arranged in concentric circles emerging from a centre point to build up the roof structure. The roof itself was made from aluminium sheets, giving the dramatic reflective sheen to a geometrically perfect roof form. A steel box section ring beam contained the outward thrust of the roof and brought loads down to the ground via a series of lenticular metal inclined struts.

With a change of government in 1952, large parts of the Festival site – including the Dome of Discovery – were cleared. In position for just a matter of months, the quality of design, the technical expertise, the sci-fi imagery, and the pure novelty of this building provided inspiration and remained a strong influence to successive generations of engineers and architects.

Norman Foster – one of the most significant architects with whom Hunt was to work in later years – has referred on many occasions to science-fiction, Space Age adventure and even Dan Dare as a source of inspiration. He commissioned the former *Eagle*'s graphic artist John Batchelor to recreate sci-fi imagery for the Renault Distribution Centre building at Swindon. This was included as a part of the Architectural Review's feature on the building in 1983. He also made reference to Flash Gordon's 'blade of light' in the publicity surrounding the Millennium footbridge in London. The futurisic imagery provided by science-fiction has continued to fascinate designers, especially when it represents freedom in personalised transport, and associations of adventure. The year 1984 saw the Los Angeles Olympic Games and its opening ceremony, where the centrepiece was the arrival of 'rocket man' – an expression of personal liberation made possible by technology. This spectacle, available to the eyes of the world, realised the science fiction dreams of the immediate post WWII era. As technology has continued to develop possibilities to fulfil these imaginings, several objects besides the rocket backpack predicted in a science-fiction future have been realised. In 2008 (as predicted in the sci-fi writings of Michael Moorcock) the use of the human body as an airframe powered for winged flight was spectacularly demonstrated by a flight across the English Channel by Yves Rossy, using his home made jet-powered back-mounted wing. The wing does not include normal moving parts such as flaps or rudder. Rossy controls flight by moving his head and by adjusting the balance of his weight.

Tony Hunt had a season ticket to the Festival of Britain and was impressed by the structures there, especially Skylon (a landmark mast intended to provide a visual reference for the Festival), designed by architects Powell and Moya and which Felix Samuely had ingeniously engineered.

Organisers of the 1951 Festival of Britain required a 'vertical feature' as a landmark element for the Southbank site. Architects Powell and Moya won the competition for its design. Their design proposed a reflective and luminescent vertically-positioned cigar shaped mast. Nearly 80 metres long and pointed at each end, it measured 91 metres to its tip and yet was suspended 12 metres above the ground. Three equilaterally positioned lattice steel pylons, angled upwards and outwards from the ground provided support. Skylon was purely sculptural and had no functional use. Although appearing cylindrical, Skylon was actually constructed as a 12-sided steel framework in 3.6 metre sections stacked vertically. Aluminium louvered cladding panels were attached giving a reflective sheen to the structure.

Felix Samuely – who had tutored Hidalgo Moya when he studied at the Architectural Association – helped devise the post-tensioned steel structure that held the mast in position. Three sets of twin cables anchored to the ground were directed to the tip of the angled pylons. At this point the twin cables split, with one strand supporting the base of the mast and the other connected to the mast half way up. The unsupported upper part of the Skylon structure was therefore cantilevered from its midpoint upwards. Samuely's ingenuity was evident in his use of jacks below each of the three pylons. Adjustment of these jacks provided the method for the post-tensioning of the cables, thus stabilising the structure.

Tungsten lamps within the framed structure illuminated the louvered cladding and made the Skylon mast glow at night. The design, with its tension structure that was invisible at a distance, gave an impression of Skylon floating above the ground without any visible means of support. The resulting sculpture portrayed an evocative image, perhaps as an imaginary, as yet undesigned space rocket. It fired the imagination of the UK population, who were eagerly anticipating a future science-fiction world.

The Festival of Britain site was laid out for large numbers of visitors, and contained several terraced areas where visitors could rest. Of particular note were the outside chairs supplied in large numbers. For this purpose Ernest Race (1913–1964) designed and manufactured the 'antelope chair' – a lightweight wire construction with balled feet. Race's BA3 chair, made from recycled recast scrap aluminium from the aircraft industry, and the above-mentioned antelope chair won gold and silver medals (respectively) at the prestigious 10th Milan Triennale in 1954. Although born in Gosforth, Newcastle-upon-Tyne, Ernest Race had been educated at St Paul's School in London before going on to study interior design at the Bartlett School of Architecture. The BA3 chair was originally exhibited at the 'Britain can make it' exhibition held in London in 1946.

In 2008 publicity from engineers W.S. Atkins was given to a project to recreate Skylon in its original form. This proposal appears to have stalled, but indicates a recurring theme in design since the mid 1980s: the desire to recreate strong images of design from the past, rather than develop a new technological statement for the future. The reassurance and provenance of a familiar symbol from the past would find a wider acceptance than a bold technologically-advanced venture of unknown pedigree.

The Royal Festival Hall was the only principal building from the Festival of Britain to survive beyond 1952. It might be considered incongruous that in design terms the buildings, furniture, fabrics and many other designs have stood the test of time, whereas another principal exhibit – a steam railway locomotive – would anchor the designs on show to an earlier period altogether.

Hunt's interest in lightweight materials and engineering prompted him to seek employment with F.J. Samuely and Partners. Born in Vienna in 1902, Felix Samuely was an Austrian Jew who had moved to the UK and away from his homeland in the early 1930s to avoid unrest. He was in the United States at the time of Hunt's employment interview and so Hunt was interviewed and then offered a job by Samuely's deputy, Frank Newby (1926–2001). Frank Newby, who married Evelyn Hogg, proved to be a much valued friend and colleague to Tony Hunt throughout the early years of his career. At the age of 32 Newby became Senior Partner at F.J. Samuely and Partners (this upon the death of Felix Samuely in 1959) – a position he held for 32 years. Newby died in 2001. In many ways Tony Hunt's career followed a similar path to that of Newby. Both were leading engineers of their generation, both were Fellows and Gold Medalists of the Institution of Structural Engineers, and both were awarded Honorary Fellowships (FRIBA) of the Royal Institute of British Architects. Neither one was particularly interested in status for its own sake; they were not committee men, but when an interesting design challenge was put before them, both would respond enthusiastically with a continuous flow of ideas – often new – but always responding to the architectural needs facing them.

Tony Hunt's appointment to work at F.J. Samuely and Partners was the turning point in his early career. He stayed with F.J. Samuely and Partners for seven years and he describes his time there as having 'taught him a lot of what he knows today … learnt in that amazing free spirit office'. Whilst he was there, the Samuely office moved from its original location of Hamilton Place just off Park Lane to nearby Dover Street (the office was actually in Grafton Street, on a corner, which confusingly allowed the use of Dover Street as the address) in the Mayfair area of London. In Samuely's office young

Portrait of Felix Samuely – described by Tony Hunt as a 'quiet, reserved and private individual'.

GUESTS

Sir Ove Arup
J.M. Austin-Smith CBE
John Brandon-Jones
John Broadbent
Peter Dunican
Dr. Wilem Frischmann
Birkin Haward
Anthony Hunt
Frank Newby
Colin Penn
Sir Philip Powell OBE

FELIX SAMUELY DINNER

MENU

Melon Boats

Veal Escalopes
in Rosebud Sauce

Sauté Potatoes

Haricots Verts

Chocolate Mousse

Coffee

AA WINES AND LIQUERS

Red Wine	£2.00 per bottle
White Wine	£2.00 per bottle
Malmsey Madeira	75p per glass
Armagnac VSOP	75p per glass
8 Year Old Malt Whiskey	75p per glass

Please order the above from the waitresses.

Thursday 25th March 1982 at 8.00pm

[Right] *The menu for a dinner in honour of Felix Samuely, with a list of engineering colleagues including Anthony Hunt.*

[Below] *A contemporary black and white photograph of Skylon whilst under construction.*

engineers were encouraged to collaborate with architects in design and here Hunt discovered what could be achieved by this bi-disciplinary approach to design. This sense of co-operation had been fostered by the Architectural Association and also by Ove Arup, who had seen this as a way forward in his dealings with Berthold Lubetkin and the Tecton group of architects.

Hunt commenced work for the Samuely practice and elected to join a group headed by one of the partners there, Sven Rindl. On the day that Hunt started work, Sven was absent through illness, leaving Hunt to fend for himself. The most significant project that Hunt worked on at Samuely's was the United States Embassy building in Grosvenor Square, London. He worked on the precast concrete structure for the upper floors. Design for the embassy building was by the USA-based Finnish architect Eero Saarinen – a Modernist. Also, Hunt worked on lightweight composite concrete and steel structures for school buildings in Staffordshire and Warwickshire. Construction for these roofs comprised a series of lipped concrete planks, which became a permanent shuttering for an in-situ reinforced concrete slab above.

Another project that Tony Hunt remembers from his time at Samuely's was the design for a pre-stressed timber-folded roof for the St Clement Dane comprehensive school in Chalfont St Giles, completed in 1953. Here, in a bold move, Samuely elected to use balloon wire for the pre-stressed tension cables.

[Left] *Modernist concrete structure of the USA Embassy in Grosvenor square, London, designed by architect Eero Saarinen. (Image courtesy of Andy Matthews)*

[Below] *The Bald Eagle – the symbol of the USA – positioned centrally atop the Samuely-designed precast concrete wall panels.*

Hunt recalls that Felix Samuely himself was a quiet, reserved and private individual (yet firm in his opinion), which – Hunt says – may appear odd in that he was able to foster such a relaxed yet energetic office environment. Hunt recalled one occasion when Wilem Frischmann, having spent many weeks developing a reinforced concrete design for the US Embassy, was simply told by Samuely: 'Oh, that's too big, make it half that size.' At that stage most calculation was quite laborious, carried out using a Monroe calculator.

Felix Samuely is credited with the structural analysis of the interconnected wafer-thin spiral ramps of the Modern Movement masterpiece of design, the Penguin Pool at London Zoo, designed by architect Berthold Lubetkin and the Tecton group of architects of which Lubetkin was the Head. Born in Russia in 1901, Lubetkin had studied under Auguste Perret in Paris, but had moved to London in 1932. Other members of the Tecton group had graduated from the Architectural Association in London. Amongst their first commissions was the Penguin Pool for the London Zoological Society. There were also similar projects for the zoos at Whipsnade and Dudley. The design heritage of the London Zoological Society at their Regent's Park site dates back to work by Decimus Burton in 1827 when the gardens were first laid out.

Ove Arup is also credited with the concrete design for the Penguin Pool. At the time both Arup and Felix Samuely were working as concrete technology specialists with the contracting firm J.L. Kier & Co. (Felix Samuely worked for J.L. Kier & Co. for just a few months in 1935). What follows is some contextual history about Ove Arup's past and his connections with Felix Samuely. Born in Newcastle-upon-Tyne (but of Danish descent), Ove Arup had moved from Christiani and Nielson to J.L. Kier & Co. (both companies with Danish origins) in 1934 when their offices were moved from Stoke-on-Trent to London – a move to attract Ove Arup to join them. Arup moved because he sought greater bi-disciplinary cooperation between architects and engineers, and whereas Christiani and Nielson were experienced in civil engineering (mainly for the maritime and harbour constructions), they did not offer sufficient support to Ove Arup for his architectural ambitions. After leaving J.L. Kier & Co. in

1935 after just a year with the company, Felix Samuely set up independently and kept close contacts with a number of Modernist architects who had moved from Mainland Europe to the UK prior to the outbreak of WWII. This group included historian Nikolaus Pevsner, architect Ernö Goldfinger and for a time before he moved to the USA, architect Walter Gropius. The MARS Group (Modern Architecture Research) – active from 1933–1957 – was the UK group affiliated to CIAM (Congres Internationaux d'Architecture Moderne) – a forward-thinking group of architects founded by Le Corbusier and Walter Gropius and others. The MARS Group included many of the associates of Samuely: Serge Chermeyeff, Berthold Lubetkin and other members of the Tecton group. Incidentally, there is reference in Peter Jones' biography of Ove Arup of a CIAM meeting in Hertfordshire and London in 1951 (1951 is a key date in the Tony Hunt story). The organising committee and speakers included Walter Gropius, Le Corbusier, Ernesto Rogers (Richard Rogers's cousin), and Serge Chermayeff. This shows that the continuing commitment to Modern Movement design and politics survived beyond WWII.

One further contextual digression worth mentioning is that Felix Samuely had worked with architect Serge Chermeyeff on the De la Warr pavilion at Bexhill-on-Sea. Herbrand Sackville – a socialist Peer and 9th Earl De La Warr – became Bexhill's first socialist Mayor in 1932. He invited entrants to a competition for the design of a sea front pavilion, which was opened in 1935. Out of over 200 competition submissions, the design was awarded to architects Eric Mendelsohn (who had left Germany to work in the UK when Hitler came to power in 1933) and Russian-born Serge Chermayeff (who had already established his design reputation). To complete the design team Viennese-born Felix Samuely was appointed engineer to the project. Drawing upon this wider European expertise, the De la Warr Pavilion became a masterpiece of Modernism, providing a restaurant, auditorium, a library and reading room, and what was to become its defining feature: roof terracing for promenading and sunbathing. The upper levels were accessed by a sweeping round staircase, the tower for which was a projected cylindrical glass enclosure. All surfaces were designed for maximum light transmission and reflection, with full height slender steel-framed glazing set into white rendered walls, and all above polished, reflective terrazzo floors. Felix Samuely brought to the De la Warr project what – up to the time of this building – had been an idea developed in Germany of a site-welded steel-framed structure. The pavilion used lattice steel beams fabricated with straight bottom booms, but with angled top booms giving the pitch to the traditional looking roof over the auditorium. Felix Samuely had produced for exhibition before the MARS group a 'Plan for London' in association with another European Modernist architect, German born Arthur Korn. Felix Samuely – with his associations to the Modern Movement architects – was aware of the political affiliations of the MARS Group. The De la Warr Pavillion had socialist connotations. Whilst he was working with Ove Arup at J.L. Kier & Co. Samuely would have known some of the leading architects of the movement, including Berthold Lubetkin and other members of the Tecton group. Projects from 1935 – Highpoint Flats in Highgate and Finsbury Health Centre in London – epitomised the Modern Movement ideals both politically and architecturally.

The Penguin Pool is of an overall elliptical shape, encompassing a long pool and a deep tank (diving pool). The outstanding feature spanning over the pool is a pair of wafer-thin spiral interlocking ramps that have no intermediate support. The Penguin Pool and their ramps were designed to allow the penguins to behave naturally and called for an innovative solution to what was then a very unusual subject for engineering design. The pool continued to be used for the next 70 years or so, but in 2004 the current generation of penguins was moved to a new building at the zoo.

As well as their engineering activities, the offices of F.J. Samuely accepted architects working towards their professional qualifications, thereby providing work experience (albeit in an engineering environment). Of the architectural students he encountered at this time, Hunt refers to architect and critic Patrick Hodgkinson – who became a lifelong friend – and Katrina Chollerton, who would work closely with Frank Newby. John Prizeman and Robert Byng were architects who also trained at Samuely's offices.

Whilst working at F.J. Samuely and Partners (Samuely's), Tony Hunt married Patricia Daniels. They moved into 39 Gloucester Crescent in London. They had two children – a son Julian born in 1959 and a daughter Polly born in 1961. Julian Hunt became an architect and furniture designer and Polly became a textiles specialist.

Julian (b.1959) and Polly (b.1961) as children photographed in a cottage garden.

Tony Hunt was developing what was to become a lifelong passion for industrial design as well as structural engineering. In 1959 he entered the Cantu International Furniture Design competition (in Italy). His entry received an 'honourable mention'. Buoyed up by this success, Hunt left Samuely's to join Terence Conran at his firm in North End Road, London as a designer. This partnership only lasted six months and ended when Hunt fell out with another partner there.

Hunt recalls an episode when he was still at Terence Conran's: Conran had designed an exhibition stand for a company named Newman Hender – a valve manufacturer. Together with a female colleague, Christina, Hunt transported, erected and helped to attend the stand in Hanover for the week's duration of the exhibition. In some way, this success of Hunt's involvement with the stand contracting operation in Hanover, and the personal relationship he built up with Christiana Smith seemed to aggravate jealousies within the company. On Hunt's return, he became embroiled in a row with one of the other partners at the firm, which signalled the end of his association with Terence Conran.

In 1960 Hunt became the in-house engineer to the firm of architects Morton Lupton (which later became Hancock Associates), specialising in timber roof structures – including the roof for the synagogue at Carmel College in Oxfordshire. This still exists as a ski slope-shaped roof finished with zinc sheeting. Tom Lupton and John Morton – after graduating from a formal architectural education – set up two firms working in parallel. One was Morton Lupton Architects and the other was L.M. Furniture Ltd. Confusingly, the furniture division of the business was Lupton Morton (L.M. Furniture Ltd), whereas the architectural side was Morton Lupton. The furniture business manufactured and sold pieces designed by Lupton and Morton from a warehouse building in Earlham Street, Covent Garden, London. The top floor of this building was to become AHA's first offices. In the end, the furniture business was not successful and encountered financial difficulties. In a twist of fate, the business was bought out by Terence Conran prior to the establishment of Habitat.

During his time at Hancock Associates (Tony Hunt was one of the associates together with Ivor Smith and John Toovey) interior design work on the new Shell Centre – built on the South Bank in London – brought Hunt into contact with Sir Hugh Casson and Lady Casson. Sir Hugh Casson was an architect and artist – head of RCA. Lady Casson (Margaret MacDonald) was an interior designer and photographer. The two were lead designers for a part of the building's interior. The Cassons were responsible for the small and large boardrooms; Tony Hunt was designer for the smaller of the two boardrooms. The architects for the Shell Building (built in 1953–1963) were Easton and Robertson who interestingly – with Partner Ralph Maynard-Smith – were early pioneers of European Modern Movement in the UK. The Easton Robertson partnership was responsible for the design of the Royal Horticultural Society (RHS) New Hall at Vincent Square, London built from 1923–1926. This building – also known as the Lawrence Hall – is a very early example of parabolic (defined below) concrete arched portal frames and stepped slab roof construction in the UK. Heating coils were contained within the concrete ceilings, giving a radiant heating system. This was an early example of integrated design between structural engineering, architecture and mechanical engineering, seen more widely with Modern Movement architecture in the 1930s. As a brief aside, Howard (Morley) Robertson (1888–1963) studied at the Architectural Association in London, and became Director there from 1920 to 1935. He was one of the first British architects to promote European Modern Movement buildings in the UK. Robertson was in Partnership with (John) Murray Easton (1889–1975) who is credited with developing and realising the practice's designs. Easton and Robertson also designed the British Pavilion at the Paris Exposition des Arts Decoratifs of 1925. Here they would have been influenced by European Modern Movement buildings especially those from Denmark and Austria. Another partner at Easton and Robertson was Ralph Maynard-Smith (1904–1964), also a graduate of the AA who went on to be an acclaimed Modern Artist.

The distinctive parabolic arches of the aforementioned Lawrence Hall (for the Royal Horticultural Society) are not truly parabolic in shape, due to the architects' vision that the columns should appear vertical and 'normal' from ground level. Hence, unlike the condition if the arches had been truly parabolic in shape (in which case all forces would have been resolved and brought to ground), the arched portal frames generate an outward thrust at the spring point of the curved section of the arch. This point coincides with a concrete ceiling slab over the side aisles. By evaluating this slab as a horizontal beam of 45 metre span and 8 metre depth, the outward thrust of each arched frame is contained. In turn, the pair of matched horizontal beams are stabilised by steel ties built within the end walls. Here, we can see an example of architectural considerations leading to compromises in the structural design. Tony Hunt was to become adept at responding to similar demands from the architects with whom he worked.

Hunt enjoyed his work for the Shell Building, and contact with the Cassons led him to a close association with the Royal College of Art. Terence Conran – Hunt's former employer during his short time away from structural engineering also lectured at the RCA. Hunt lectured there and even produced a structural design manual for their newly-founded course in architecture.

Other significant projects during Hunt's time with Hancock Associates were Jenkins Garage at Wallingford, Oxfordshire, and a group at townhouses at Priory Mews. Here, within the development, Tom Lupton and John Morton – both partners at the firm and therefore Hunt's employers – owned a property. They had set up a firm called 'Townmaker' to carry out the property development. The site for development was a part of John Morton's garden.

After the break-up and re-launch of the practice as Hancock Associates, the properties had to be sold. Hunt – who was living in one of the town houses – recalls the difficulties of partners having to resolve their differences whilst living as next-door neighbours. Once the houses had been built at Priory Mews, they did not sell quickly and so they became studio offices for Hunt and other associates of the firm.

When Tony Hunt left Hancock Associates, John Morton – a Principal Partner at Hancocks – took legal action against Hunt over a structural survey Hunt had carried out on behalf of the company. The case concerned the much heralded Barley Mow Workbase. This was a new concept of the time (much copied since) of renting out workspace within a fully serviced and staffed building. The architect and also one of the developers for this project was David Rock. Tony Hunt acknowledges that as a result of an incomplete structural survey (in which the fact that structural beams were assumed to be steel when in fact they were wrought iron), Hancock Associates were able to press their claims for negligence. This claim amounted to little more than a personal vendetta carried against Hunt by John Morton. Hunt reflects that Tom Lupton had little or no part in the action. Hancock Associates themselves were not being claimed against, so it is evident that the action was purely personal. In mitigation, Hunt explained that provision for the necessary inspection access was woefully inadequate. The outcome of this dark period – at which time Tony Hunt was setting up his new practice Anthony Hunt Associates – was an out-of-court settlement that Hunt financed by appointing partners David Hemmings and John Austin to the new Anthony Hunt Associates practice. Tony Hunt tells us that the first he knew of this problem was a phone call from David Rock suggesting that he 'checked his professional indemnity insurance'. The claim amounted to some £250,000.00, but thanks to some skilled negotiations from Hunt's accountant Brian Humphrey, a settlement of £90,000.00 to be paid over a three-year period was agreed. However, Hunt explains that good can come from the most unlikely set of circumstances, and that the result was a solidly-founded partnership that was to thrive and prosper from that point on.

After his stint of lecturing on structural design for the architecture/interior design courses, Tony Hunt continued for several years to lecture at RCA for one day each week on the subject of structures, and specifically for the course in furniture design.

His fascination with the design processes for furniture continued to provide parallels in the design of components for building structures.

Hunt also tells us that: 'Much further back, in the early days of AHA, I worked with Mary Shand [architect Jim Sterling's wife] when she worked for architects Russell Hodgson & Leigh'. Together they worked on the Coylum Bridge Hotel at Aviemore; she as interior designer and he as engineer. He was also engineer for the ice rink roof there. Similar to Carmel College, the ice rink roof comprised bulky gluelam beams of compound shapes, supported by substantial square columns. There is a striking silhouetted photograph of Tony Hunt acting as site engineer for the ice rink roof, perched high in the air with only a flimsy ladder for access, inspecting a structural connection. Russell, Hodgson & Leigh were an architectural practice specialising in interiors. Sir Sydney Gordon Russell (1892–1980) – the founding partner – was an acclaimed furniture designer who lived and worked in Broadway, Worcestershire in the Cotswolds. He was responsible for the design and manufacturing technology of WWII 'utility furniture': a component-based, batch-produced, low cost furniture range. Tony Hunt met Gordon Russell on a few occasions, but he knew and worked with Mary Shand of the practice and Dick Russell – Gordon's son. A project (in about 1963) of historical interest was the restoration, structural support and exhibition of the ancient Greek Nereid Monument, located in Room 17 of the British Museum. Pieces of the monument (built as a tomb for Lycian leader Erbinna c.390– 380 BC) had been brought from Turkey to London in the 1840s by British traveller Charles Fellows. Parts of the tomb had been stored for many years in the basement of the British Museum, and under Russell, Hodgson and Leigh's direction, a concrete supporting structure was designed by AHA. In addition, key components such as the Ionic columns and capitols, were replicated as structural elements where originals had been lost. Tony Hunt describes the setting up of an on-site grit blasting plant in a temporary tent within the galleries of the British Museum. The newly cast concrete parts were eroded by grit blasting so that they resembled the original pieces. Figures of Nereids are featured in relief panel sculptures located between the columns. Nereids are mythical sea nymphs.

Tony Hunt remembers Robin and Lucienne Day – specifically their work at the Festival of Britain – both for interior design and furniture design. He was aware of Robin Day's work (though they did not meet), but it was Lucienne Day that Tony Hunt knew, having met her at various architects' parties in the Covent Garden area of London. He describes Lucienne as 'fascinating and charismatic'. The Days' most notable work at the RCA was before Hunt's time there. During Hunt's career there are constant encounters with graduates of either the Architectural Association or of the Royal College of Art, or with designers who spent their formative years in the offices of architects who had associations with these institutions. Whether Tony Hunt sought out these like-minded people, or whether these emerging creative talents needed Hunt to give structure to their architectural visions, is not quite clear; it is probably a combination of the two, where the talents of Tony Hunt complemented their own. One such encounter was with Yorkshire-born Jon Wealleans (b.1946) – a friend of Tony

[Above] *Nereid Monument of Ionic Order, recreated in room 17 of the British Museum.*

[Left] *Structural connections for timber beams receiving Tony Hunt's full attention.*

Anthony Hunt Associates

Hunt's who had graduated from RCA after spending just a term at the Architectural Association School of Architecture. He had worked briefly for Foster Associates but in 1969 'Mr Freedom' (a shop selling fashion items and furniture) opened at 430 Kings Road, Chelsea. Wealleans-designed furniture was for sale in the shop. He went on to establish the architectural practice of Kemp Muir Wealleans. At that time, Hunt didn't have the chance to meet Keith Critchlow – an architect and lecturer at the RCA with whom he was to work later on the Crestone Dome project – even though they were both present at RCA. However a book, published in 1969 by Keith Crtichlow entitled *Order in Space* influenced Tony Hunt. The book is described as a design source and examines the geometry of patterns and grids, 'where mathematics and arts meet' – an echo of Ralph Tubbs's description of 'mathematical poetry' in his design of the Dome of Discovery for the Festival of Britain in 1951. Buckminster Fuller was a mutual friend of both Tony Hunt and Keith Critchlow. Other books that had an effect on Tony Hunt at about this time were *Turning Point of Building, Structure and Design* by Konrad Wachsmann published in 1961; and also *Philosophy of Structures* by Eduardo Torroja (1899–1961) – a Spanish engineer specialising in concrete shell design: this book was published in 1958 and Hunt refers to this book as holiday reading during a trip to Ireland.

Tony Hunt cites two other Europeans engineers as having had an influence upon his design philosophy. Firstly the Spanish engineer Felix Candela (1910–1997) who designed thin shells of reinforced concrete. Secondly, the Italian engineer Pier Luigi Nervi (1891–1979) who specialised in concrete structures, notably the Sports Palace in Rome built for the 1960 Olympic Games. Both were awarded Institution of Structural Engineers Gold Medals: Candela in 1960 and Nervi in 1967. In addition, Z.S. Makowski's 1965 book *Steel Space Structures* was an important influence on Hunt's own vision of structural design. Another book that Hunt mentions is *Functional Tradition in Early Industrial Building*. This was jointly authored by Eric de Maré (1910–2002) – who studied at the Architectural Association but specialised in architectural photography – and J.M. Richards (1907–1992) – a renowned writer on architecture. With this 1958 book, industrial architecture became worthy of study and exposed lattice steelwork together with plain unadorned concrete structures took their place in the architect's vocabulary.

Descriptions of grids and patterns as a generator of lightweight structural steel design is evident in the work of Fritz Haller, both in his steel building and USM furniture systems. Tony Hunt is the proud owner of an untitled book (beautifully printed), detailing all of Fritz Haller's work, presented to him personally by Haller. The application of appropriate structural grids, based upon strict geometric patterns became a feature of Tony Hunt's work. He explains that it is the structural connections between various members that occupies his thought

Tony Hunt is currently planning to move house again, and by coincidence the house he is planning to buy belongs to the aged widow of Eric de Maré.

during the design process, but these connections are ultimately dictated to by their alignment to a grid. Identification and resolution of suitable and appropriate structural grids became essential in the development of the High-Tech movement. Repeated use of components, with the possibilities that prefabrication and batch production brought, were only relevant if the structural grid provided the means for modular construction.

One further noteworthy point is that during his stint of lecturing at the RCA, Hunt was aware of James Dyson – a student of his – and his industrial design work. Dyson moved directly from the RCA to work on developing a business manufacturing the Sea Truck – an inshore landing craft type of boat made from moulded GRP. This is an indication of the energy and enterprise derived from experience gained at the RCA – that an arts student should have the confidence to enter a world of engineering.

3   HUNT'S HISTORICAL LINKS

The history of lightweight structural engineering owes much to the aircraft industry. During WWII new manufacturing techniques and materials were introduced. Richard Buckminster Fuller envisaged factory manufacture of his Wichita House to solve the twin problems of what to do with the aircraft industry capacity when they converted to peaceable activity immediately after WWII, and to address the housing shortage in the post war austerity with cost effective supply. His big idea was to harness the latent capacity and intensity of wartime manufacture for the wider good of society.

It was the assembly line at the Beech Aircraft Corporation in Wichita, Kansas USA, where Buckminster Fuller proposed to produce his Wichita House. It is worth noting the differences of opinion between Buckminster Fuller and Ove Arup – the latter of whom might be considered to have been a voice for the wider socialist and political views of the Modern Movement. Fuller thought that technology could solve the problems of society. Arup insisted that technology would only ever re-arrange physical elements – only human will and resources could solve the problems.

Comparable conditions in the UK produced a similar response for housing and school construction. But the relevance of the aircraft industry story in the UK pre-dates WWII and takes us back to the late 1920s, when Neville Barnes Wallis was working for Vickers and he designed the R100 airship in Howden, Yorkshire. The lighter than air principle of hydrogen-filled gas bags held within gigantic lightweight structures demanded innovative engineering and manufacture. The innovation was twofold. Firstly, new processes for aluminium tube-making gave exciting new possibilities for lightweight structures. Secondly, the airship hangars themselves – also the factory to make the airships – literally took lattice steel portal frame structures to new heights.

It was Barnes Wallis who first coined the phrase 'geodesic' for his aircraft structures. Geodesic very soon became automatically associated with Buckminster Fuller's dome structures. Geodesic is taken to denote the line of the shortest possible route between two points on a curved surface: in Buckminster Fuller's case a dome; but in Barnes Wallis' case a tubular aircraft fuselage. During WWII Barnes Wallis designed the Lancaster Bomber, where he used this structural principle. It is said that Barnes Wallis was 'Victorian' in his engineering approach – pioneering a new transport industry and suiting new buildings to the new demands similar to the establishment of the railway industry in the second half of the 19th Century.

The Howden Airship Station first opened in 1916. The twin sheds each 229 metres long 55 metres wide and 43 metres high – built in 1919 and subsequently modified – were the location for the manufacture of the R100 airship designed by Sir Barnes Wallis. This airship made a successful flight to Montreal, Canada and back in 1930. Barnes Wallis had worked for Vickers in Barrow on an earlier airship – the R80 – which made its last flight to Pulham Airship Station on September 20 1921. Whilst working for Vickers in 1923, he had conceived the idea of large-scale rigid frame construction for 'lighter than air' airships, but had not been able to put the theory into practice until the R100 project. He had also worked on ideas for mooring masts, which allowed the huge airships to slew around the masts and face into the wind. The Airship Guarantee Company – for whom Barnes Wallis worked – was the Vickers privately-funded company responsible for the design and manufacture of the R100 on behalf of the UK government. The so called Empire Communications Scheme envisaged coverage of outlying Empire locations for mail and passenger services. The proposed routes would include Canada, Egypt and India (as well as the UK), and facilities were built in these locations in readiness for the service.

One of the twin bays of the hangar at Howden in effect became a template for the construction of the R100. The enormous 12-sided polygonal cross-section frames of the airframe, adjoined by longitudinal ribs running the length of the airship, were hoisted into position and suspended from rails in the roof of the hangar. Here they remained until construction was complete. At this stage, hydrogen filled gas bags were inflated by hydrogen gas ducted into the hangar from pipelines under the floor. Generation equipment housed in ancillary buildings supplied the hydrogen. Thus the hangar became an integral part of the design methodology. The R100 measured 216 metres in length and 41 metres in diameter. A clearance of less than 1 metre remained between the airship and the hangar walls as it floated out through sliding doors at the end of the hangar. It became the means to construct the R100 as well as its home. There was no mooring mast at Howden, so the craft had to return to the hangar after each flight, or be moored to the ground using a tetrahedral arrangement of guying cables. The hangar was constructed from buttressed portal frames of lattice steel, clad with corrugated steel sheets. The helical winding of lightweight duralumin strip with an underpinned seam jointing system used to make the tubular truss components (from which the polygonal frames for the R100 were made) signified a technological breakthrough. Barnes Wallis' assistant engineer and stress calculator on the R100 between 1926 and 1929 was Nevil Shute Norway (who became the celebrated novelist). His structural calculations took many months to complete, particularly in respect of the sizes of tensioning cables used to stabilise the polygonal frames. These cables were designed to remain in tension even when the airship was subjected to dynamic loads if manoeuvring or lifting.

When Barnes Wallis left the project to work for Vickers Aviation in Weybridge, Surrey, Norway became Chief Engineer at Howden and took part in most of its flights. He remained in this position when the R100 was taken to the government operating station at Cardington in 1929. In October 1930, following the abandonment of the whole airship project, the Airship Station at Howden closed and was allowed to fall into disrepair. Little evidence remains of the activities today but Tony Hunt visited the sheds at Cardington in the 1970s when the Building Research Station (BRS) used one of the two sheds there as a test station. He may even have seen and been inspired by the Cardington mooring mast as he travelled north from London. This remained as a symbol to the great airship enterprise until its demolition in 1943, when Hunt would have been 11 years old. Also, in around 1982–1983 Tony Hunt recalls that with his architect friend Dominic Michaelis (they shared an office in Paddington at the time), he visited the hangars at Cardington. In the empty hangar (that one not used by BRS), Michaelis and Hunt carried out a test flight of Michaelis's hot air balloon. They successfully laid out the balloon on the floor of the hangar before inflating the canopy, firing up the burners and lifting off. The planned exit flight through the hangar doors, however, was discreetly abandoned.

By 1929 the Cardington Airship Station was set to become the centre for international airship operations with its hub in the UK. An essential part of this strategy was the 62-metre steel-framed open mooring mast. Measuring 21 metres in diameter at its base and the first cantilevered tower of its type, it enabled the 'lighter than air' airships to swing freely around the mast and face into the wind. They could be safely moored in this way. Weights were attached to the tail end of the craft to complete the mooring operation. By use of a winch mechanism, the airships could be brought safely into dock at the mast head. The necessity for exact positioning reliant upon the airship itself (in particular any reversing manoeuvre) was thus avoided. The mast had a staircase climbing up the centre of the tower allowing passengers and service personnel to access the airship via a high-level landing platform. Even so, a steeply angled gangplank was required to climb from the top of the tower to the airship's passenger cabin. Hydrogen from a generating plant and a water supply were designed to be available at the mast head. The winch house – which accommodated the winding gear – was a brick built structure at the base of the mast.

Cardington was the only UK location for the government-backed Empire Communication Scheme, but sites across the world were selected. A mast similar to that at Cardington was constructed near Montreal in Canada. This was used when the R100 airship travelled across the Atlantic in 1930. On this occasion, one of the flight crew descended by parachute to supervise the docking of the airship. However, facilities in Egypt and India that had been built in readiness for flights of the R100 and R101 were never used. The Cardington mast was finally demolished in 1943. Barnes Wallis and the R100 can be seen as a 'half-way house' in the look back to the Victorian era and the monumental engineering successes of that period. Both Barnes Wallis and I.K. Brunel are cited as strong influences on Tony Hunt. James Dyson confirms that he remembers the impact that images and descriptions of Brunel's life and work had upon him: this as a result of Hunt's lectures at the RCA. I.K. Brunel became an inspiration to Dyson.

Tony Hunt tutored James Dyson in the development of the Sea Truck – a project that Dyson undertook whilst he was still a student at RCA. This project was the start of a co-operation between Dyson and Jeremy Fry (1924–2005). Fry was a renowned engineer (he was a friend of Sir Alec Issigonis, 1906–1988) and arts patron. He had studied at the Architectural Association in London and was the son of Cecil Fry – the last chairman of the famous Fry's chocolate company.

Similarly, Joseph Paxton's work at Chatsworth House and at the Great Exhibition building of 1851 pioneered the repeat use of prefabricated standardised components – a theme that Hunt was to embrace. Joseph Paxton – Head Gardener at Chatsworth House – pioneered the use of prefabrication in the construction of large-span glass-houses, anticipating the methods he was later to use on the Great Exhibition Building of 1851. Sir Joseph Paxton (1803–1865) became Head Gardener at Chatsworth House, Derbyshire in 1826 at the age of 23. He had impressed William Cavendish, 6th Duke of Devonshire, whose London residence Chiswick House was close to Paxton's place of work, Chiswick Gardens. In the first half of the 19th Century there was a new-found interest in exotic plants brought back from tropical regions. Such plants required hothouses. In 1837 work started on the Great Conservatory at Chatsworth House, where the maze now stands. Decimus Burton – the architect who later worked with ironworker Richard Turner on the Palm House at Kew – drew up plans for a building 84 metres long and 37 metres wide. The double-vaulted cloister design – where one curved roof section sat on top of another – gave an internal height of 19 metres over a central carriageway running the length of the conservatory. Ridge and furrow glazing in conjunction with the orientation of the glasshouse maximised sunlight transmission by allowing morning and evening sunlight (when at its least penetrative) to strike the glass surface at right angles. This method had the added advantage of allowing flat glass to be formed into a curved vaulted roof; this by tessellating rhomboid shapes. Coincidentally, this gave an organic, patterned, leafy structure to the glazing, which resonated with the plants on show. The Chance Family glass company in Smethwick provided 1.2 metre lengths of glass, whereas up to this point only 0.9 metre lengths had been possible. The same company went on to produce glass for Paxton's Great Exhibition building (1851) and for the Houses of Parliament building (1840–1860). The ridge and furrow method of glazing also used straight sash bars – unlike the Kew Palm House where moulded glass was used to match the curved iron glazing bars. Important as the design of the Great Conservatory proved to be, Paxton also broke new ground in the technological approach to the building process. He utilised steam-powered woodworking machines to produce standardised components in a way that had not been seen before. Iron columns and accessories were used, but arches, glazing bars and gutters were built up from timber sections. It was this systemised approach that was to emerge as the means to build the Great Exhibition building, and Paxton envisaged its use in London on an even larger scale in the unbuilt scheme, The Great Victorian Way of 1855.

The Great Conservatory could be described as the first building in Britain to engender an industrial aesthetic – a product arising directly out of the use of a multiplicity of batch-produced components. This building proved Paxton's technical and organisational credentials. In 1849 a second conservatory was built at Chatsworth House. The Lily House was created to house a single species of lily – Victoria Regia – recently brought from the Amazon and transferred from Kew Gardens where it had failed to thrive. This building was Paxton's own design, (as opposed to the jointly designed Great Conservatory, for which the design credit must be divided between Paxton and Burton), in which he still used a combination of wood, iron and glass. He developed a flat-roof version of the ridge and furrow glazing method, and a curtain wall system of hanging vertical bays of glass panelling from cantilevered beams. It was the combination of these techniques, the glazing methods and the batch production of components that led the way to the design of the Great Exhibition building. Paxton continued as Head Gardener at Chatsworth House despite becoming a Member of Parliament and a leading figure in the emerging railway industry. He retired when the 6th Duke died in 1858. The Great Conservatory and the Lily House were demolished in the early 1920s after falling into disrepair.

Similarities can be found in the engineering approach of Joseph Paxton and Barnes Wallis. They both tailored new manufacturing methods to suit their vision of the technology required at that time in history. Paxton introduced steam-powered woodworking machinery to mass produce components that he needed to complete his vision. Neither the methods nor the components had existed up until that point. Barnes Wallis developed a technology for the helical winding of aluminium strip (called duralumin) and a method of riveting the seams from below to produce a smooth finish that would not tear the fabric of the airships. From this technique he was able to manufacture lightweight tubes from which lattice beams were made. These beams were in turn made into the enormous air frame of an airship. It was the novelty of the manufacturing methods that gave rise to the advance in technology necessary for the realisation of the visions of both Paxton – in his designs for the Great Conservatory and Lily House at Chatsworth House – and Barnes Wallis in his designs for the R100 airship. Tony Hunt was able to bring a fresh understanding of technology to structural design, and whereas the technology wasn't necessarily new, its transfer from other industries to building often was, and thus his application bears comparison with Paxton or Barnes Wallis.

Working for Felix Samuely gave Hunt direct lineage back to Modern Movement ideals, both aesthetically and politically. Samuely – who was able to impart such passion for design to Hunt – had personal and professional contact early in his career in the UK with Lubetkin, Ove Arup and others who were to shape British Modernism. The advances in concrete technology both practically and structurally, were to determine the ethos of the Modern Movement, just as timber woodworking had determined Paxton's work, and aluminium tube making had determined that of Barnes Wallis.

The apartment blocks Highpoint I and Highpoint II (which was completed a few years later) in the Highgate area of London are considered masterpieces of the Modern Movement. The concrete structure responds to the architectural demands both functionally and formally. Highpoint I was built for Tecton's client Sigmund Gestetner, who was interested in providing housing for his workforce. His business, based in the Camden area of London, was the manufacture of well-known office equipment. To the Tecton group and the Modern Movement, this was an ideal set of circumstances: an informed client who subscribed to the Modernist ideas, with a sympathetic outlook on contemporary culture. Gestetner's willingness to appoint Tecton added support to their left-wing political views, which promoted socialism and community values. Lubetkin made claims that Highpoint would be 'for working class families'. The phase I building contains 64 flats in two seven-storey blocks connected on plan into two cruciforms. Entrances to all the flats are through a common lobby that contains a winter garden and at a lower level, tearooms and a terrace. There are also shared rooftop terraces designed for community living – a powerful theme derived from the work of Le Corbusier and others in the Modern Movement. Modern Movement émigré architects Erich Mendelsohn (1887–1953) and Ernö Goldfinger (1902–1987) and their families each became tenants of Gestetner's in Highpoint I. The conceptual community aspect of the development can be seen to have been realised with the bringing up of families within the apartment structure. Mendelsohn's daughter, Esther was a regular babysitter for the Goldfinger children. Goldfinger was one of the first tenants for Highpoint I, where he stayed for a time before building his own much admired Modernist housing terrace in Willow Road, Hampstead. Mendelsohn moved to the USA.

Advanced features of Highpoint I included the use of removable platforms to construct the in-situ shuttered concrete walling and cantilevered balconies. This – a technique derived from marine concrete constructions – avoided the need for scaffolding as the platforms were reassembled at a higher level each time a 'lift' was required. Other innovations included ceiling-mounted hot water radiant heating panels, built-in refrigerators with a communal compressor in the basement, two main passenger lifts but separate service elevators for kitchen use, and a system of built-in cabinets and wardrobes with roller-shutter fronts.

Ove Arup himself was critical of the concrete design, which he called '[a] muddy kind of structure'. His opinion was formed because of the difficulty in analysis of the concrete structure. Walling panels were considered at that time to have little or no intrinsic strength and calculations had to assume a frame structure carrying the in-fill concrete walls. The result was an over supply of reinforcing steel where columns and beams might be. Concrete was required to be poured in a wet condition to flow into the shuttering at these points. Similarities can be seen with Owen Williams's Dollis Hill Synagogue: here Williams sought to use the introduction of odd-shaped openings (some hexagonal, some inverted round arch shaped) as a reason to define the stress patterns of the walling.

Highpoint II is an extension to the adjoining Highpoint I but in the event, only three years later than Highpoint I, its design is very different to its predecessor. There had been local opposition to the design concepts of Highpoint I and under the powers of the Town and Country Planning Act of 1932 local residents were able to dictate fundamental changes to the design. Highpoint is the highest point in London and the site's prominence brought close attention from the public to its design. The building was restricted to one fifth of the size of Highpoint I, and the development ended up as 12 luxury maisonettes instead of the 57 workers' houses as Lubetkin and Gestetner originally intended.

The design took the form of rectilinear blocks extending from one leg of the cruciform blocks of Highpoint I. It comprised two conjoined eight-storey elements again with a shared lobby entrance and a roof terrace. The end elevations of the extensions contained two maisonettes per floor which continue the facade design of Highpoint I. The central zones are differently configured, with recessed two-storey glass ribbons (of steel framed glazing) and a cantilevered balcony on every other floor. The concrete end walls were built using the same techniques of shuttering as had been proved in phase I. However the central area has a concrete frame with brick spandrel infill between floors. The staircase towers are expressed as vertical bands of glass blocks, which start their ascent above ground level, leaving the band of glass blocks appearing to 'float' unsupported above the ground.

Such was the success of the Modernist idea of high-rise living, that architect Berthold and his wife Margaret Lubetkin lived in a personalised penthouse rooftop apartment at Highpoint II from 1937 until 1955, before moving to the countryside and Lubetkin distancing himself from architecture to become a farmer. As properties, Highpoint I and II have remained much sought after ever since. Both blocks are preserved as Grade I listed buildings. It represents a twist of fate that a scheme originally devised as utilitarian workers' housing with Modern Movement credentials should have become an exclusive piece of real estate over the decades. This about-turn started with an abandonment of the socialist principles that were the generator for Highpoint I, in favour of exclusivity by Lubetkin as he designed Highpoint II.

Expertise in concrete technology brought several other Tecton/Modern Movement buildings, such as the Gorilla House (1934) and Penguin Pool (1935) at London Zoo, and the Finsbury Health Centre building (1939). Central to this group was a socialist belief in equality and a determination to make under-privileged Finsbury – then the second smallest of the London Boroughs – a model of social progress in housing, education and health. Lubetkin's message at the opening in 1938 was that 'nothing was too good for ordinary people' – he thought that architecture should be an instrument of social change. Lubetkin and the Tecton group of architects had already worked on socially aware community-based building schemes such as Highpoint I – a development intended as 'workers' housing' for the employees of Sigmund Gestetner, the office equipment manufacturer. Indian doctor Chuni Katial commissioned the Finsbury Health Centre on behalf of the Public Health Committee of the Borough Council of Finsbury. It was opened in 1938 by the King's Physician, Lord Holder, and has remained open as a health centre ever since.

The Finsbury Health Centre building is of symmetrical layout: H-shaped on plan with two wings flanking a central block. The central block contains a reception and waiting area and used to have a so-called 'electrical treatment' – in fact an area to provide artificial sunlight. Above is a lecture theatre and a part-covered terrace. Splayed wings of the building funnel visitors toward the bronze framed, fully glazed entrance doors, which are set into a wall made from glass blocks. Due to its southwest orientation any sunlight is captured into the grand foyer, which has hard reflective surfaces (and used to have red painted columns to reflect its political hue). Lubetkin called this 'a happy optimistic place'. Originally furniture by Alvar Aalto adorned the waiting areas and the reception area was an oval of polished marble. In a similar vein to Highpoint I planning, structure and services were ingeniously integrated. Patients need not climb stairs for their treatment – all clinics were located on the ground floor. The Modern Movement influences – both political and architectural – were never more evident than in the design of this building. Floors in the wings of the H plan spanned between the long walls. These were cast as sandwich slabs with hollow spacing blocks between two layers of in-situ concrete. Heating coils were cast into the lower of the slabs to provide ceiling heating to the rooms below. The floor slabs were supported by concrete channels formed in-situ as continuous horizontal service ducts. The top flange of the channel acted as a continuous windowsill for the ribbons of opening steel-framed windows (on a teak sub-frame). Removable spandrel panels were used as facing to the concrete channel, allowing services to be accessed from outside. Concrete mullions supported the concrete channels and the whole structure was stabilised by concrete end walls to each block. The building is now Grade I listed, as are other buildings by Lubetkin/Tecton including the Regents Park Zoo Penguin Pool, Highpoint I and Highpoint II.

In addition to the historical links, Tony Hunt had forged new links with key engineers and architects of his own generation. Konrad Wachsmann – the American architect engineer – worked during WWII on aircraft hangars. Wachsmann himself had links back to the Modern Movement. He worked with architect Walter Gropius (one of the founding members of the CIAM group of Modern Movement architects) on a prefabricated timber housing system for the General Panel Corporation in 1941–1942. It was Wachsmann's Open Hangar system of 1950 – a space frame structure using the Mobilar system with its possibilities for flexible layouts – that provided inspiration for Hunt and the architects with whom he was working. Walter Gropius, born in Germany in 1883, brought with him influences of the Bauhaus Movement (in 1925–1926 he had planned a new Bauhaus complex in Dessau). Gropius moved firstly to the UK in 1934 where he worked with architect Maxwell Fry, and then in the early 1940s he moved to the USA. There he worked with Marcel Breuer and became Professor of Architecture at Harvard University. Wachsmann was trained not in structures or in building systems manufacture, but as a cabinetmaker. He was able to transfer the practical skills thus learned to component and structural design in much the same way as Tony Hunt (from an engineering training) was later to emulate. Wachsmann was a close friend of industrial designers Charles and Ray Eames. He became Director of Building Science at the University of Southern California (USC) in Los Angeles. The Department of

Architecture at USC had Charles Eames and Craig Ellwood as visiting lecturers and they, together with several other protagonists of the case Study Houses programme, had associations with USC. Architect Pierre Koenig – a graduate of USC – designed Case Study Houses Nos. 21 and 22. Tony Hunt refers to the Case Study Houses as fine examples of lightweight, component-based building design. Wachsmann was a leading authority on building prefabrication – he established the then-only centre for research in industrialised construction at USC. In 1953–1954 Wachsmann produced an image known as 'Vinegrape' of a structure of apparently infinite dimensions, all created from the repeat use of a single component. Judith Wachsmann drew the image. In Europe, Mero Structures and its founder Max Mengeringhousen were developing space-frame structural systems along similar lines. The 1961 Khartoum exhibition provided early evidence of the external prismatic trussed frames that would become central to the High-Tech Movement.

In the UK, the Arcon prefabricated (prefab) house epitomised many of the factory-built prefabrication mass housing ideals for post WWII house building. Ove Arup – more often associated with Modern Movement concrete architecture – joined a wide range of specialists for design work on the project. Although only a modest steel structural frame was required, the co-ordination of component design, manufacture and supply proved to be very successful.

Towards the end of WWII, in response to a severe shortage of housing in the war-torn towns and cities of the UK, the Ministry of Works sought designs for factory-produced prefabricated houses that could be delivered and erected quickly in any location. In July 1944 a shortlist of three such designs –including the Arcon House – were demonstrated at the Tate Gallery in London so that a design could be selected for future use. WWII had not yet ended in February 1944 when Ove Arup became a member of the Government Prefabrication Committee. By the summer of 1944, bombing had created half a million homeless citizens in London alone.

Designs for the Arcon Mark V were developed by architect (George) Edric Neel (1914–1952), who 'passionately believed that architecture should work with industry'. In 1943 Neel formed the Arcon Group with Raglan Squire (1912–2004) and Rodney Thomas (1902–1996) – former associates at Raglan Squire Architects. Arcon was a then-new type of research organisation sponsored by several manufacturers with the aim of improving links between industry and architecture. Edric Neel had previously worked for the Modernist architects Wells Coates and Denys Lasdun (Lasdun was a junior member of the Tecton group of architects headed be Berthold Lubetkin). Orders were placed for considerable quantities as up to 150,000 such houses were anticipated by the Ministry of Works. The second most popular prefabricated house at this time was the Aluminium Bungalow B2. In fact it was the Arcon Mark V that became the standard house type. With overall plan dimensions of 7.45m x 7.16 m, it came complete with bathroom and kitchen, and boasted such refinements as ducted warm air heating, modular kitchen fittings, pre-wiring using electrical cable harnesses, and prefabricated floor and ceiling panels. The house stood on a simple concrete raft slab.

The supply of these houses became a huge undertaking. 41,000 were actually built – less than expected but even so, 145 manufacturing companies were involved and 5000 working and component drawings were produced. There were some 2500 items and components for each house. The last Arcon houses were delivered in 1964. What was intended as a temporary solution to an urgent housing shortage became a mass-produced design, examples of which have lasted for many decades. Arcon itself continued until 1967 continuing to research into materials and techniques for prefabricated buildings, especially for tropical countries. As a research organisation, Arcon never designed buildings to individual commissions and after Neel's death the organisation lacked the impetus he had created. Ultimately Arcon was absorbed into the Taylor Woodrow construction group at Taywood Engineering, Greenford West London.

There were other post-WWII images that would have had an impact on Tony Hunt besides the Arcon House. The Bailey Bridge was a temporary bridge structure made up from a mass-produced standard kit of parts, which could be configured to suit a wide variety of spans. The Mulberry Harbours, central to the WWII Normandy Landings, would be another powerful image of 'engineering in action' from this period.

In October 1943 the Ministry of Defence instructed some 550 engineering contractors in the UK to build various elements for the two now famous Mulberry Harbours (Harbour A for the American-led army and Harbour B for the British-led army) that were required to enable the Normandy Landings to take place on Omahah Beach and Arromanches respectively. It was an audacious manoeuvre pivotal to the outcome of WWII. All component elements of the harbours were designed and built from scratch in just a year. At the planning stages of the project, Winston Churchill famously said: 'bring me the best solutions, do not waste time with the problems, they will take care of themselves.' Although all design work was carried out under the terms of strict secrecy, most of the acclaimed engineers of the time – consultant engineers Oscar Faber, Ralph Freeman, Ove Arup and Alan Harris, and contractors Malcolm McAlpine and Norman Wates – were involved, bringing their expertise to the project.

Also, there was the ubiquitous Nissen Hut. This standardised building was of very simple design and was capable of rapid construction. It was used for a whole variety of purposes including classrooms, dormitories, administration offices or workshops.

Jean Prouvé's Maison Tropique of 1949 was designed to be air freighted in the hold of an Air France plane of the time. (Image courtesy of Iqbal Aalam)

In France, Jean Prouvé used a practical hands-on approach to the design of lightweight buildings. He envisaged buildings as a kit of parts. His *Maison Tropique* of 1949 was designed to be air-freighted in the hold of an Air France aeroplane of the time. Here is an early example of High-Tech design responding to extremes in environment – in this case the extreme high temperature of French colonial Africa. Emphasis was placed on solar shading and natural ventilation. Aluminium components are derived from the aircraft industry. Tony Hunt met Prouvé during his work with Richard and Su Rogers up to the point of the establishment of Piano and Rogers. Prouvé was a member of the judging panel and was instrumental in the award of the project and their appointment as architects for the Pompidou Centre at Beaubourg, Paris.

4 THE SIXTIES ADVENTURE

London in the 1960s became the focus for a new culture of self-expression in fashion, music, theatre, cinema, art and design. This search for a liberated lifestyle spilled over into architecture and engineering. After the austerity of the immediate post-WWII period of the 1950s, a new world of architecture, engineering and megastructure was imagined: Utopia.

It was to be just a matter of time before a solution would be found – an inevitable conclusion of the machinations of this generation of visionaries. The 1960s was a fertile period for dreams of architectural solutions to the ills of society, whether they were to combat housing shortages or whether the buildings were for the purpose of fun. Which Utopia should we choose? Would it be a New Babylon with streets in the sky, spreading wherever our wishes took us? Or would it be a walking city that kept on moving until an ideal location was found? If conditions changed, we simply walked on. Or would Utopia be under a domed glass roof where climate was controlled and complete cities continue to function below? Or would it be a high tower with views above the clouds or on floating man-made islands offshore? Or indeed, Utopia may be a travelling experience without any fixed location, cruising on a luxury liner or an airship with restaurants moored to a New York skyscraper. Or would Utopia be a science-fiction world of colonies on the moon, below the ground or in space? Architectural and engineering design has always played its part in this dreamworld – it takes the mind's eye of the visionary filmmaker, artist, architect or engineer to realise the images of a science-fiction world. The sixties generation of architects inherited from industry, fashion, film, and pop culture a design momentum that has carried them through to the present day.

Projects such as the Milllenium Dome (now the O2 Arena, designed by Richard Rogers Partnership), its temporary next-door neighbour Skyscape (designed by ESS, itself the closest manifestation of Cedric Price's Fun Palace) and the Eden Project may show considerable technological advances, but derive their imagery from the 1960s era. Most of the film and fashion statements of 1960s London revolved around the idea of fun. It was fun to dodge the London traffic and negotiate the historic streets in a brightly coloured Mini, and fun to move freely in town or country on a Moulton bicycle. Incidentally, Eleanor Bron (b.1934) in her 1978 book *Life and Other Punctures* describes her travels through Europe riding a Moulton bicycle, nicely linking this engineering icon into her world of comedy, theatre and cinema (Eleanor Bron was married to Cedric Price, the visionary architect). The book *Homo Ludens: A Study of the Element of Play in Culture* (published in 1938 by author Johann Huizinga [1872–1945]) portrays Man as a fun-loving animal who lived to enjoy himself.

The British arm of the Pop Art Movement is considered to have been launched by the 'This is Tomorrow' exhibition in 1956, under the auspices of the Institute of Contemporary Arts (ICA) – a group active in London from 1952–55. Theo Crosby – the then-editor of *Architectural Design* – arranged the exhibition. It included essays by Reyner Banham and the exhibition guide was by renowned architect Colin St John Wilson. Other members of the group included Peter Smithson – architect and year tutor at the Architectural Association (who in 1956 contributed designs for House of the Future at the Ideal Home exhibition) – and Pop Art artist Richard Hamilton. It was Hamilton who wrote to the Group setting out what Pop Art was about. This letter included a list of attributes that the movement might include: popular (designed for a mass audience); transient (short term solution); expendable (easily forgotten); low-cost; mass-produced; young (aimed at youth); witty; sexy; gimmicky; glamourous; big-business. In 1965 Cedric Price applied most of Hamilton's epithets in his 'Pop-up Parliament' project: 'popular', 'transient', 'expendable', 'witty' and 'gimmicky' all match Price's vision of that time. He called the resulting design 'a supermarket for politics'.

At the Architectural Association, Archigram (under the stewardship of Peter Cook and other members – both staff and students) and also Cedric Price (working closely with Frank Newby) provided the seeds of change from which the High-Tech Movement emerged. Cedric Price studied and lectured at the Architectural Association in London, but although contemporary with the Archigram initiative, he was not directly involved with their designs. Price developed projects which in a small part shared imagery with Archigram, but in the main, his were more politically driven, more realistic in the choice of sites and functions, and more aware of interaction within neighbourhoods. He derived his imagery from transport and industrial process. Price worked on photographic prints by superimposing images of industrial equipment, structure and vehicles. He was more interested in how the assemblies fitted into their local surroundings than he was in their appearance. He saw buildings as temporary assemblies of industrial equipment placed strategically into areas already identifiable and established, whether it be as an industrial wasteland or existing parkland. He attempted to weave his architecture into the existing fabric of the locality, transforming its identity without inflicting damage. Price saw his architectural assemblies as having a limited (and probably short) lifespan: there for as long as they were useful in much the same way as industrial plant. Cedric Price only suggested the appearance of his buildings and little attention was paid to their 'buildability', or to what materials might be used for their construction.

The 'Potteries Thinkbelt' – Cedric Price's 1966 project – was centred upon the disused railway and coal mining areas of Staffordshire in the UK. He saw the 'Thinkbelt' project as a replacement for two outdated industries in which an infrastructure was established with its associated employment, to provide mechanised education services. Cranes, gantries and railways were taken from traditional industry and their images transposed to new architectural forms. Cranes lifted, gantries supported and railways transported, providing mechanised man-handling to get people to the right place at the right time for education.

The idea that architecture should be fun was a reflection of influences from other areas of popular culture, including fashion, art, cinema and music. It was formally presented in a number of schematic architectural projects of the sixties. The 1968 Beatles film *Yellow Submarine* reflected a backdrop of 1960s culture – an exercise in fun. Animated graphics by Czech born Heinz Edelman – illustrator for the German youth magazine *Twen* – took the place of actors and location filming. Bright colours and simplified forms alluded to an uncomplicated world. The graphic sequence accompanying the song *Sea of Holes* is a derivation from the work of Victor Vasarley (1908–1997), the Hungarian-born French abstract painter and European master of Op-art. The imagery of *Yellow Submarine* bears a resemblance to the Archigram images of 'Instant City' (Peter Cook 1969), where an airship takes the place of the submarine. The term 'psychedelic' originally referred to the effect of hallucinogenic drugs, but now has become associated with the Endeman style of graphics that are brightly coloured two-dimensional representations of simplified abstract forms. Op-art patterning and tessellation techniques are used to stimulate a heightened sense of perception (which had originally been induced by drugs). This graphic style was used in the drawings of Pop-art interiors by Ralph Adron (b.1913) – a theatre designer. 'Take a Room' for *The Daily Telegraph* in 1968, as envisaged by Max Clendenning (b.1934) – an architect, interior and furniture designer – but drawn by Adron, show how the Endelman style of graphics was used in the representation of an architectural interior. In this case the centre perspective was reduced to the minimum to maintain a two-dimensional image, which accorded with the Endelman graphic style. This alliance between architect and theatre director which gave rise to futuristic architectural forms – as demonstrated by Clendenning and Adron – had already worked a number of years earlier in 1959 when Joan Littlewood (1914–2002) had asked for 'some young nuts' (her words) to produce a design for a site at Lea Valley in the east end of London. Cedric Price and Frank Newby answered this call and produced images for the Fun Palace. Littlewood wished the design to reflect and recreate the identity of Vauxhall Gardens – a long lost area of London, dedicated to fun and leisurely pursuits. It was here in 1836 that Charles Green successfully undertook a flight in a balloon which took him to Nassau in Germany. Promenades, orchestras and tearooms afforded the leisure pursuits of that time.

As an aside, in 1969 Peter Cook of Archigram used imagery of airships in designs for Instant City and in 1964, Ron Herron of the Archigram group published science-fiction cartoon-style images of Walking City. It is also worth mentioning that, as a student, Mike Webb – a founding member of the Archigram group – had produced and re-worked an innovative project 'Sin Centre' (1958–1962). Unlike Cedric Price, whose projects did not give much idea of what the building might look like, Webb's project was developed to a more advanced stage, with a geometrically complicated glass roof. This project pre-dated Price's Fun Palace, realistically addressing problems of structure, construction and servicing. And, from 1958 through until the 1970s, the Dutch painter Constant A Nieuwenhuys (1920–2005, known simply as Constant) presented a long-running series of images called 'New Babylon'. His visionary concept was for an infinitely expandable modular city structure dedicated to the concept of fun: a place where leisure and pleasure had taken over from work as the mainstay of life.

Joan Littlewood (1914–2002) was a visionary theatre director working in Stratford, East London. She held passionate views concerning the role of theatre as a positive political and community influence. She said: 'theatre should be free like air, water and love.' She promoted a revolutionary concept of community theatre, to which architect Cedric Price and engineer Frank Newby (from F.J. Samuely and Partners' office) – both lecturers at the Architectural Association in London in the 1960s – were able to give physical form.

The Fun Palace – originally thought of in 1961 – would cover a large area of 29,640 square metres, being 260 metres long and 114 metres wide. Resembling a shipyard or container terminal, giant gantry cranes would dominate the structure and would be used to locate, access and position various plug-in modules. Service towers would have fixed locations together with a grid-like lattice beam and column structure, into which the unitised modular building sections could be placed. Even the escalators – the means of access to the theatres and workshops – would be capable of movement, allowing complete flexibility of layout. Roofs would be retractable, only to be used when and where required, and the most unusual feature was to have been the absence of a fixed floor. All modules would be suspended from the structural grid. The whole assembly would be suspended above landscaped parkland.

Cedric Price famously claimed that time was the fourth dimension of architectural design and advocated buildings which could expand, retract, reconfigure and adapt to the changing demands. Frank Newby was sympathetic to his views and devised a structure of lattice beams and columns as a skeleton onto which floors, walls, bridge links, roofs, ductwork and services could be attached, and which would be looked after by the mighty gantry cranes spanning every corner of the enormous building. It was intended for a short life of 20 years and may never appear in the same configuration on more than one occasion – the Fun Palace was there to stage events, not as a monument to traditional theatre. Its layout, form and identity would be generated by the mass of consumers which used and surrounded the project – not imposed upon it by the preconceived ideas of traditional patrons. This design concept influenced British architects and engineers and it was the Pompidou Centre built in Paris (1972–1977) by architects Piano and Rogers and engineers Ove Arup and Partners that most closely resembles the design of the Fun Palace. The Lloyds of London building (by

Richard Rogers Partnership, again with engineers Ove Arup and Partners) repeats some of the features stylistically. Rooftop cranes there give a glimpse of the potential ease with which the building could be adapted in the way that the engineers and architects of the Fun Palace had advocated.

In 1960, Lord Snowdon undertook the creation of the aviary, which now bears his name. The architect he appointed was Cedric Price, the visionary architect (who lectured at the Architectural Association), and as engineer he sought the expertise of Frank Newby (who had only recently taken on the leadership of F.J. Samuely and Partners after Felix Samuely died in 1959). The rich design heritage of London Zoo dates back to 1827, when Decimus Burton (1800–1881) – the architect responsible for the Palm House at Kew Gardens and the Great Conservatory at Chatsworth House – first laid out the grounds and designed buildings for the Zoological Society of London. The Snowdon Aviary at London Zoo was one of a very few buildings actually built by Cedric Price. From F.J. Samuely's office, Frank Newby and Wilem Frischman worked on the structure. This project passed through the F.J. Samuely office in readiness for its construction in 1960, just after Tony Hunt had left. But as Hunt continued to keep close contacts with Frank Newby and others at the practice, he will have seen and no doubt commented on the design proposals for the aviary.

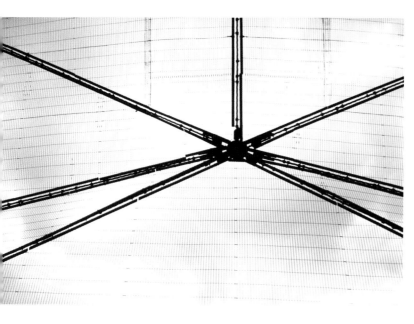

[Right] *The mesh enclosure is supported by a tensile structure, anchored in turn by compression tubular corner posts.*

This design team produced a structure exhibiting the qualities of innovation, daring, excitement and spontaneity, which had characterised the nation's spirit of optimism a decade earlier in the Dome of Discovery and Skylon at the Festival of Britain. Indeed, Felix Samuely's ingenious tension structure for the Skylon mast was a forerunner of the tension structure that Newby developed for the Snowdon Aviary.

For this quasi-building – essentially a netted enclosure – a pioneering design was demanded. Technological innovations included the use of aluminium castings, stainless steel forgings, and lightweight welded mesh to include the maximum obstacle-free volume for the birds' unencumbered flight. Assemblies of tetrahedral tubular compression structures at each corner of the rectangular building act as anchorage for longitudinal tension cables to which the netting is attached. The roof – a pair of crossover cables running lengthways along the apex of the aviary – is held in tension, supported by tubular gable post columns formed into giant v-shapes. The use of tubular compression members and tension cables is a clear expression of the cable net structure.

As mentioned above, in the 1960s Cedric Price had advocated buildings that could adapt and change according to circumstance, and need not be permanent. How ironic it is that one of the few schemes of his ever to be built should form a part of the proud heritage of design and engineering of Zoological Society of London dating back nearly two centuries.

The 1960s was a time of experiment, daring and adventure. There was a perceived need for a newness in architecture and engineering, leaving the past behind and embracing an unfettered future. In some ways Tony Hunt had the best of both worlds: his experience with Samuely had given him expertise in the Modern Movement style for concrete walls, floors, columns and beams, and his wider design interests had provided background in lightweight materials and manufacture which might provide the building fabric of the future. In other words, in addition to a traditional means of construction, Hunt was able to promote a bolt-together, dry form of prefabrication and by design he was able to avoid wasteful and thus expensive forms of construction. Hunt has often said that he enjoyed designing concrete structures.

The Modernist architects specialising in concrete technology also played a part in the sixties adventure. The Smithfield Poultry Market with its clear span concrete shell roof is testament to the willingness of its architects T.P. Bennett and Son, working with Ronald Jenkins and Povl Ahm of Ove Arup and Partners to develop innovative designs trusting to advances in technology. This display of mastery of pre-stressed concrete technology belonged to the era of the 1960s when 'everything was possible' and there was a real sense of optimism, adventure and derring-do in architecture and engineering. This combination of knowhow, ambition and expertise has rarely been seen since in a concrete shell roof construction.

Hunt himself lived on a houseboat for a time, not in the aforementioned area but still on the Thames at Wallingford. He says this was an unsuccessful enterprise resulting in the sinking of the boat on more than one occasion. The boat had been bought on a whim by Hunt's former employer John Morton, who had failed to have a survey carried out before buying it.

Having acquired his considerable experience in the analysis of lightweight structures, Tony Hunt was uniquely positioned to provide professional credibility to what were – at that stage – vividly displayed but untested structural concepts. Most of the futuristic images were really only two-dimensional and often, Tony Hunt had little more to go on than cartoon-like graphic images. He relished the opportunity to realise the new architecture in sound reliable structural forms.

In August 1962 Tony Hunt founded Anthony Hunt Associates. The move was made partly out of necessity, as his employment with Hancock Associates had ended abruptly in July of that year. The years working for F.J. Samuely and Partners had provided Hunt with continuing good relations with Frank Newby. It was Newby who provided subcontract work for Anthony Hunt Associates in the formative months of independent practice. The history of Samuely's contacts with Modern Movement architects furnished Hunt with introductions to the Architectural Association in London and with an emerging group of graduate architects.

Hunt also met with leading architect engineers from the USA and Europe, including: architect engineer and prefabrication specialist Konrad Wachsmann (whom Hunt met c.1982 in Los Angeles when he was lecturing at UCLA – Pat Hunt and he had dinner with Wachsmann and his wife in their house); industrial designer Charles Eames (c.1978 Hunt visited the Eames House case study house number 8); designer and philosopher Richard Buckminster Fuller (whom Hunt met on many occasions); architect and industrial designer Fritz Haller (Hunt eventually met Fritz Haller in Gratz in 2001); and building and industrial designer and fabricator Jean Prouvé (whom Hunt met on occasions when Prouvé visited Foster's and Rogers's London offices).

Photographer Pat Hunt photographed by husband Tony in Colorado, 1982.

Hunt had worked on the USA Embassy building in Grosvenor Square, London during his time at F.J. Samuely and Partners. Architect for the building was Eero Saarinen who, having spent much of his working life in America, had become a close friend of Charles Eames. Indeed, Saarinen had co-operated with Eames on the design of the Eames House of 1945–1949 (case study house number 8 in Chautauqua Boulevard, Pacific Palisades in the Los Angeles area of California). Saarinen also worked with Eames on the design for a neighbouring house – case study house number 9 of the same period – this time for the organiser of the case study programme, John Entenza. At that time Entenza was the editor of *Arts and Architecture* magazine who promoted the CSH programme.

Saarinen was interested in other aspects of industrial design and he designed the 'Tulip Chair' in 1956. This chair was contemporaneous with Eames' Lounge Chair and Ottoman (Eames' cast aluminium group of chairs were slightly later in 1958). The 'Tulip Chair' was made by Knoll and Associates whereas Eames worked principally for Herman Milller. This provides an example for the overlap and removal of traditional boundaries between architecture, engineering and industrial design. Hunt was to continue in this emerging discipline. It also shows a promising willingness to meet, share and exchange design and technological advances, even though Eames and Saarinen had been working for rival manufacturers.

As a noteworthy aside, Eero Saarinen moved to the USA when he was 13 years old in 1923. He grew up within the community of Cranbrook Academy of Art in Michigan where his father Eliel Saarinen taught (Eliel was an architect and it was he who designed the buildings for the Cranbrook Academy). Eero studied sculpture and furniture design at Cranbrook and became close to fellow students Charles Eames and Ray Kaiser (the future Mrs Eames), and also with Florence (Schurst) Knoll. He went on to study architecture at Yale University School of Architecture, completing his studies there in 1932. Charles Eames worked in the offices of Eliel Saarinen in 1939.

Tony Hunt also refers to Achille Castiglioni (1918–2002) as a designer that he admires. His folding table 'Cumano' of 1978–1979 provides the qualities for which Hunt himself was striving – those of lightness, colour, boldness, simplicity and yet refinement by using materials sparingly. Castiglioni received his doctorate in architecture from Milan Polytechnic in 1944 and from 1969 he taught industrial design at Turin Polytechnic.

With the benefit of these influences and jointly with a wide group of architects which included Norman Foster, Richard Rogers, Michael Hopkins and Nicholas Grimshaw, Hunt was instrumental in the pioneering of a lightweight, prefabricated, component based, industrialised style of British architecture – later to be called High-Tech. This label, 'High-Tech' was always unwelcome to the above-mentioned group of architects, and every attempt was made to shake off the name.

Hunt recalls that his first meeting with Richard Buckminster Fuller was at the house of a neighbour, Geoffrey Bocking – a lecturer at Corsham College. Buckminster Fuller gave his lectures in two parts. He spoke for up to two hours and then retired for a rest before returning to conclude his lecture. Hunt actually worked with Buckminster Fuller on a project for an underground extension to St. Peter's College, Oxford. Sadly, it was never realised. With Foster Associates the scheme was for an authorised Samuel Beckett Theatre built underground, below St Peter's College. Tony Hunt describes the scheme as a vaulted hemi-cylindrical underground construction with hemi-spherical end sections.

Hunt also refers to James Mellor, who had adopted the role of Buckminster Fuller's agent in the UK. Mellor had worked at Foster Associates with Jan Kaplicky (who was a founder of Future Systems) – they left Foster at the same time to go their separate ways. Jan Kaplicky – born in Czechoslovakia in 1937 – founded Future Systems in 1979 with David Nixon. As had been the case with Cedric Price, Future Systems supplied countless images of futuristic concepts intended for mobile or temporary architectural solutions, without building or realising any of their designs. Tony Hunt had been friendly with Jan Kaplicky from the time he had moved to the UK. He moved from Prague to London in 1968. Kaplicky had also worked for a time with Piano and Rogers on the design of the Pompidou Centre in Paris. James Mellor and Tony Hunt had more indirect contact with Charles Eames at the time when the Eameses created some footage of film magnifying an image of planet earth time after time until the view changed from cosmic to microscopic proportion. The presentation was at the United States Embassy building in Grosvenor Square (as mentioned above, one of Hunt's first projects). The film was called *Powers of Ten* and was produced by Ray and Charles Eames for the much admired computer firm IBM in 1977.

West India Dock floating bridge – a futuristic concept designed by Future Systems and engineered by AHA.

Unlike Sir Owen Williams or Sir Ove Arup, Tony Hunt never saw himself as an architect engineer, but preferred to provide a more traditional engineering support role to architects. However, he coupled this with an identified need for industrially-designed batch-produced components which could add to the architectural quality in much the same way as industrially-designed furniture might be conceived. He also demanded input at early stages in design. Hunt saw the design process as a bilateral approach between architects and engineers. He did not take kindly to the suggestion that he should make an architect's designs work when they were presented to him as *fait accompli*. Hunt was never tempted to adopt the role of architect – as Owen Williams or Ove Arup had been in a generation previous to his own. No doubt Hunt had opportunities to act as architect, not least of which would be when he introduced James Dyson to a choice of architects for his factory at Malmesbury. He formed working relationships easily with clients and patrons, so if he had wished to it would not have been surprising if he had ventured into architecture of his own. With Hunt's advice, James Dyson eventually appointed Wilkinson Eyre as architects for his factory project.

Hunt has worked with countless architects on projects of many different architectural styles, but throughout his career he has never built his own house. He has always preferred to respond sympathetically to whatever style was brought before him. He has preferred to work and live within an older building, free from the constraints that might be upon him if he were to associate his name with new build of a particular style.

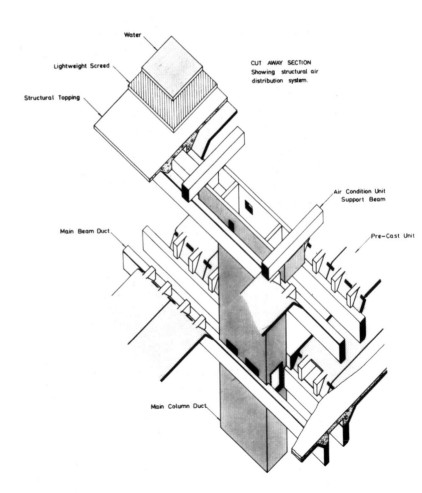

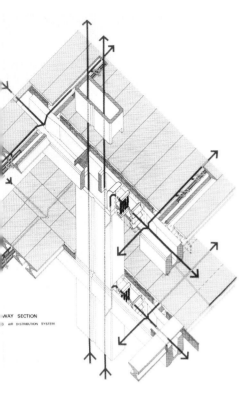

WAY SECTION
AIR DISTRIBUTION SYSTEM

With one of his first designs as Anthony Hunt Associates – Leicester University Library – Tony Hunt took the concept of integrated design a stage further, when he sought to use hollow structural concrete columns and beams for ventilation ducts. Designed by architects Castle and Park, with the structure-cum-ductwork code named 'Structair', it was ultimately intended as an off-the-peg product. This was a structure which could double as ductwork for a stack ventilation system. Unfortunately, the project was abandoned (although built subsequently to a similar design), throwing the finances of the newly-established consultancy temporarily into turmoil. It was at this time that Hunt sought the help and advice of his friend David Wolton.

Also for Castle and Park, Tony Hunt worked on student accommodation buildings at Villiers Hall, Oadby near Leicester. These comprised three-storey load-bearing masonry cross-walls supporting precast concrete floors. The concrete work was of a technique learned during his time at Samuely's: precast interlocking trough-form floor beams were laid between the load bearing cross-walls; these provided permanent shuttering for a deeper concrete slab. The roof was made from double tongue-and-groove machined timber of sufficient depth to span between the cross-walls without intermediate support. Within this development were two bigger buildings for which Hunt designed a gluelam timber beam system of support.

A project for a block of town houses at Winscombe Street in the Archway area of London brought Hunt into contact once more with Neave Brown. This was in the early days of AHA. Hunt owned a share of the Winscombe Street site off Chester

Road, involving himself for the first time in a commercial property development project there.

Another significant commission in the early stages of AHA was the Alexandra Road Housing project in north London, not completed until the late 1970s. With architect Neave Brown and working for Camden Council, Hunt engineered an uncomplicated standardised concrete cross-wall structural system, unitised to create a single terrace of housing some 300 metre long and rising to seven storeys. Neave Brown's design called for a high-rise concrete frame on anti-vibration mountings (this to counter the noise of the railway), double-glazing on the railway side, and with Max Fordham's input, heating was contained within the wall structure. Special isolators were devised between the piling and superstructure to avoid mechanical transmission through the structure of railway noise. The whole building was envisaged as a precast concrete assembly, but the contacting firm of MacInerney – totally against Hunt's wishes – dictated that the whole building would be made from poured in-situ concrete. In fact, the quality of finish was acceptable even though its components were not precast in factory conditions. In 1968, at Ronan Point (an 18 storey block of flats in the Canning Town area of London) a progressive collapse had occurred as a result of a small gas explosion within one of the flats there. At one corner of the high-rise block, an entire stack of flats collapsed. One on top of the other. The collapse was attributed to a construction method that depended upon inter-connected precast concrete panels. Analysts suggested in-situ reinforced concrete construction methods would provide structural continuity so that, in the event of one component failing, there should be no knock-on effect which might cause a progressive collapse. Bounded by a railway on the northern side of

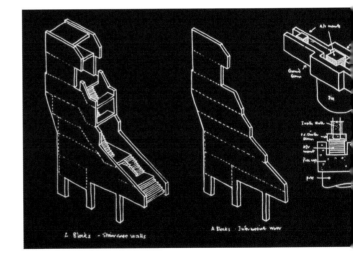

the development, the southerly-facing terraced apartment houses, with stepped balconies and raised walkways, were a fine example of high-density municipal concrete system-built housing of the 1960s and 1970s. The 512 dwellings accommodate some 1600 residents. In the 1993 the complex was given Grade II* listing. Additionally there were two other housing blocks, a community centre, and (by architects Evans and Shalev) a home for the physically handicapped and a children's reception home.

Tony Hunt's friendship and his working relationship with Neave Brown was a key episode in the early years of Anthony Hunt Associates. Hunt describes Neave Brown

[Above] *Stepped, balconied terraces of concrete structure for system-built housing at Alexandra Road, in the London Borough of Camden.*

[Below] *A view of the Alexandra Road development showing the well resolved cross-wall geometry. (Image courtesy of Jamie Barras)*

[Opposite] *Tony Hunt's sketch of the cross-wall geometry and its foundation, cushioned against mechanical sound transmission.*

The stark concrete of Alexandra Road viewed from the railway side of the development. Railway travellers will have passed this on their way into Euston Station from the north.

as a charismatic and persuasive individual, who was held in great esteem by those who knew him well, even though as Hunt says, 'he was difficult to live with'. Hunt had first met Neave Brown when Brown worked for architects Lyons Israel Ellis, and Hunt worked for F.J. Samuely. Later, Hunt and Brown shared an office at 43 Earlham Street, Covent Garden, and Hunt explains that during this time he learned some of his drafting skills and techniques from Neave Brown. Brown would draw only in pencil and favoured the use of detailing paper instead of tracing paper – the latter being more commonly used at that time. His was a relaxed, authoritative and communicative style of drawing that Tony Hunt was able to replicate. Neave Brown in turn had been taught his draughtsmanship whilst he was working for Lyons Israel Ellis. Hunt describes the drawing and design process as an exploration of concept. He quotes the Italian architect Carlo Scarpa (1906–1978): 'I want to see things, there is nothing else I can trust ... that is why I draw.' The first job on which Hunt and Brown worked together was the Post Graduate Medical Centre at Hammersmith Hospital. Tony Hunt remembers designing pre-stressed concrete beams for the roof of the building. The architects Lyons Israel Ellis – with whom Hunt came to work closely with whilst he was at Samuely's – spawned many renowned architects: Paul Castle, Alan Park, James Stirling, James Gowan, Neave Brown and John Miller to name but a few.

Tony Hunt was to work closely with several of this group of architects throughout his career. In 1956, James Stirling (1926–1988) left architects Lyons Israel Ellis and set up independently with James Gowan. Stirling became one of the leading and one of the most renowned and influential of British architects of the late 20th Century. The RIBA – Stirling Prize – a British annual prize for architecture since 1996 – was named after him. James Stirling was a visiting lecturer at Yale University at the same time as Norman Foster, Richard and Su Rogers, and Eldred Evans were studying there.

Neave Brown's time at Camden brought the engineering work on the Alexandra Road housing project to Tony Hunt's fledgling consultancy Anthony Hunt Associates. Whilst at Camden, Brown was joined by architects Gordon Benson and Alan Forsyth, who were later to set up their own practice with which Tony Hunt was to work later on the National Museum of Scotland project in Edinburgh. Neave Brown's expertise in the production of exquisite draftsmanship must have influenced Benson and Forsyth because their own practice has always promoted equally high standards of draftsmanship.

Tony Hunt acknowledges that public sector work helped fill the books in the early days of AHA. He admits that the way in which AHA were awarded work – on the basis of the most relaxed and informal interviews – just would not happen today. Neave Brown, after a short stint sharing offices with Hunt, returned to the secure employment of Camden Architects Department. From his recently acquired position of influence with Camden, Brown was able to direct operations there and appoint AHA as consulting structural engineers to what was an enormous project. Max Fordham – the building services engineer – also had a hand in the appointment of AHA for this job.

An early project that AHA took on in 1962 was the two houses he designed for architect John Gosschalk. The first, at Wallingford in Oxfordshire, was a single-storey building with exposed structure, and the second was the Dixons chairman Stanley Kalms' house – a cantilevered steel and timber-framed building.

The 1960s finished with design and engineering in the UK reaching a high point. In 1969 Concorde made its first flights, the SRN4 Mountbatten class hovercraft made its first cross-channel sailing, and the Harrier jump jet completed its trials. In the USA the Apollo missions reached their climax with the first manned moon landing. Cedric Price used images of the SRN4 hovercraft to depict futuristic urban transport in 1966 as a part of his Atom new town project. It is clear that engineering successes in other walks of life spilled over into the architectural imagery of the 1960s. These images were routinely presented to Hunt for him to give structural engineering credibility to them.

For AHA other commissions followed. The Reliance Controls factory in Swindon was a Team 4 (Rogers and Foster) project that set new standards in British industrial architecture. By using standard industrial cladding and by the uncomplicated structural connection of standard I-beams (assembled with very little notching or end preparation of sections), a building of exquisite proportions evolved. The wide span bays of uninterrupted profiled steel cladding, punctuated with exposed tensile rods for cross bracing, became a standard form of industrial building – much copied because of its logic and for its economy. There are similarities between the structural connections of Reliance Controls and the smaller timber structural connections developed by Walter Segal for his self-build programme.

However, the definitive moment for Tony Hunt was his co-operation with Foster Associates (which by then included Michael Hopkins as an associate partner) in a building for the progressive American computer company IBM at Cosham near Portsmouth. The brief called for a large single-storey open-plan temporary office. In this building Hunt engineered a large-scale yet uncomplicated economical post and lattice beam structure, prefabricated using batch-produced components. After more than 30 years of use, the designs defy any suggestion that it was ever intended as a temporary building.

Tony Hunt recalls how he got the IBM job in 1969 or 1970 when he had just taken a lease on a building in Bedford Street, Covent Garden, London. The building was much too big for his offices, but at a time of rapid expansion he was optimistic that the space there would be utilised. Hunt recalls: 'It was at this time that Norman Foster, who had by then set out in practice on his own, was to be interviewed by a man called Gerry Deighton for a so-called temporary building for IBM in Cosham. I was already in Bedford Street, and Foster's offices were still in their office/flat in Hampstead. We set up a fake office for the interview, borrowed a whole lot of furniture from Zeev Aram, made a very successful presentation and jointly got the job'. Gerry Deighton later joined architects Michael Aukett Associates.

As a relevant aside about Zeev Aram, in 1966 he had opened a showroom at 57 Kings Road, London, introducing furniture designs from Marcel Breuer and Achille Castiglioni. He had studied at London's Central School of Arts, having enrolled there in 1957. In 1966 he introduced his own designs together with the well-known designs (LC3 black leather and chrome armchair and the LC4 chaise longue) from Le Corbusier, P. Jeanneret and Charlotte Perriand. In 1973 Aram moved to 3 Kean Street in Covent Garden. In 1975 Aram struck a deal with Eileen Grey to produce her designs and in 1992 he was made an honorary Fellow at the Royal College of Art.

Originally the IBM Cosham building was intended as a temporary location for the company's UK headquarters as they awaited the construction of their permanent base. The brief called for a large open-plan office space to be built quickly (in fact construction took twelve months) – a brief that could have been fulfilled by use of a timber-panelled sectional proprietary system building. It should provide working accommodation for 750 people, possibly increasing to 1000, and associated car parking space for 660 cars. As history shows, the project became the keynote building in establishing the early reputations of its architects Foster Associates and of Anthony Hunt Associates – its engineers. Clients such as IBM with their progressive working practices and strong product identity were seen as important to the hopes and aspirations of the emerging High-Tech architectural and engineering movement.

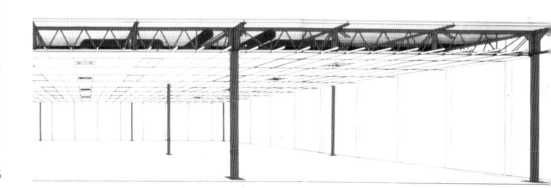

IBM Cosham is a logically planned double square single-storey open-plan building measuring 146 metres by 73 metres with a floor to ceiling height of 3.25 metres. The structure is built on a master grid of 7.3 metres by 7.3 metres supported by slender square hollow-section steel columns 125mm square. The lightweight structure rests on a thin raft of in-situ cast concrete, the design of which avoided the need for expensive piling for foundations on a poor load bearing site. The 7.3 metre square structural grid is subdivided into 7.3 metres by 2.4 metres by secondary beams of similar depth. The beams used (Metsec beams) were a proprietary product comprising a top and bottom cold-rolled member with a web of concertina rods between them. These were adapted by introducing into the longitudinal beams additional posts at junctions with the secondary beams and at column connections. The structure was speedily and economically erected by a small team using a forklift truck working off the floor slab (cast in advance of the steel erection process). The use of a forklift truck with its implications of economy and using easily handled small-scale components was a theme that Hunt returned to for future projects, including the Patera reconfigurable building.

The design solution for IBM Cosham provided for deep plan office accommodation and for reasons of traffic noise control, sealed windows were selected (these built up from a wall of bronze reflective glass supported by aluminium glazing mullions). As a result, air conditioning was required and a system was developed to serve individualised locally adjustable units to serve 7.3 metre square bays instead of the more usual centrally positioned plant room approach. The 125mm square steel columns

[Above] *Post-and-beam structure assembled by a small team of steel erectors using a forklift truck operating off the pre-prepared concrete slab.*

[Opposite] *Foster Associates' cross-sectional centre-perspective presentation drawings for the IBM Pilot HQ building.*

doubled as ductwork to carry power and telephone cables from the ceiling zone to working areas – this in the days when cables were bulky but not too numerous, before the advent of the desk-top computer when cabling became much more complex. The lattice beams were used as supporting structure for services (air, power etc) – a principle pioneered in the USA and UK for school buildings. Foster Associates and AHA had already developed much of the design in an architectural competition for a comprehensive school in Newport, South Wales. One feature of the Newport design was the proposed use of glass ceilings to separate and contain the service zones within the structural depth of the grid of lattice beams.

Before going into more detail about the school in Newport, it is worth mentioning that there was another antecedent for the IBM building: a building for Computer Technology in Hemel Hemstead, Hertfordshire. Iann Barron (the friend of Hunt's who was later to become the client for the Inmos building) ran Computer Technology at that time. With Foster Associates as architects, Hunt designed a steel post-and-lattice-beam structure for a building modest in scale when compared to IBM.

Much of the imagery and design concept for the aforementioned Newport design was attributable to Foster's time at Yale University School of Art and Architecture, where he studied under Paul Rudolf (the Head). As we know, Richard and Su Rogers were also at Yale at the same time, as was Eldred Evans. James Stirling, who tutored Eldred Evans at Yale was on the judging panel for the Newport school design competition. Hunt also identifies the work of Craig Elwood and other architects on the West Coast

Originally conceived as a temporary solution, the low cost lightweight building has been in constant use for over 35 years.

of the USA as propagating the design patterns that were adopted by Foster and Rogers and introduced to the UK. In the USA, Ezra Ehrenkrantz had developed the SCSD school building system for Californian schools. Ehrenkrantz had worked in the UK in 1954, as a scholar for the Building Research Station (BRS). Here, he would have been aware of the Hertfordshire Schools building programme – a modular system building approach to school design. Ehrenkrantz's cross-sectional centre perspective became the established pattern for many scheme drawings of single-storey lightweight buildings, such as schools and offices.

Hunt was responsible for the engineering work for no fewer than six entries in the design competition for Newport High School. Brian Forster and David Tasker – after their work on the Sobell Pavilions at London Zoo – went on to work on the Newport High School project. The successful design by Evans and Shalev was a concrete structure in the style of their earlier modernist designs, such as the ancillary buildings at Alexandra Road. This project showed Hunt's ability to respond to different architectural concepts with appropriate structural designs. Tony Hunt had become friendly with Eldred Evans through another friend – John Howard, who was lecturing at the Architectural Association. Eldred Evans was still a student at the AA at the time. John Howard had given Tony Hunt his first real job – Crescent Road Flats at Highgate Spinney in north London.

In 2008 Hunt was asked by Eldred Evans to survey the structure of Newport High School, because the local authority planned to demolish the school building, rebuild the school on a neighbouring site and turn over the present school site for housing development. Amidst the protests against the proposal for demolition, Hunt found that although there had been little – if any – maintenance, generally the structure was in good shape. Hunt obtained a copy of a structural report (critical of the state of the structure) prepared by the local authority, which had not been made available to Eldred Evans. She has been leading a campaign to prevent demolition.

Quite apart from the system-built prefabricated designs that Foster and Rogers had encountered in the USA, Swiss architect and industrial designer Fritz Haller (b.1924) was developing very similar component-based school building systems in Switzerland. Haller was able to take his concepts much further by his association with USM – a steel component manufacturer. Haller was naturally the long-retained architect for the USM factory itself, built in four phases between 1962 and 1994. Fritz Haller developed a range of prefabricated steel building systems known as 'Stahlbausystem Mini', 'Midi', and 'Maxi'. The three structural systems – all comprising lightweight lattice or lightened beams and slender columns – were ranged in size according to their use. Stahlbausystem Mini might be used for houses, Midi for schools or offices, and Maxi for factories. Multi-storey versions of the system are designed within all but the Stahlbausystem Mini. These systems were gradually developed in the 1970s and 1980s.

Fritz Haller also designed the USM 'mobel-bausystem'. This is a component-based office, studio or showroom furniture system capable of a multitude of different

configurations to suit storage or display units or workstation construction. Haller was commissioned by USM to design the furniture system in 1960 and the concept has remained in production for over 40 years. It is this product that became the mainstay of USM's production, rather than the building structures mentioned above. Due credit should be given to the mutually beneficial business arrangements to both the designer – Fritz Haller – and the manufacturer USM. Theirs has been a symbiotic relationship where each party has benefitted from the endeavours of the other. Tony Hunt was never in the position to take advantage of such a relationship with a manufacturer. In the USA, Charles Eames had become closely associated with manufacturer Herman Miller for the exploitation of his designs. By his designs, Fritz Haller achieved many of the goals that Tony Hunt had set for himself during the early part of his career. Hunt had always seen component design of structures as essentially the same as that for furniture. He considered that structures should command the same level of refinement and quality as if the components were intended for furniture – it was simply a matter of scale.

It is interesting to note that USM Haller did not remain in the business of manu-facturing and selling off-the-peg structural building systems for long. They retreated to their core business of furniture systems manufacture. This could be indicative of the difficulties in business terms of selling a standard architectural solution into this market. The USM furniture system is a stick-form rectilinear framing system with precisely-formed jointing nodes. Tony Hunt is generous in his praise of the design of jointing block used in the USM mobel-bausystem. It allows connections in x, y, and z axes and yet is sized not to exceed the dimensions of the slenderest of incoming frame mem-bers. Tony Hunt eventually met Fritz Haller in 2001 at a conference in Gratz, Austria. The meeting was at the invitation of Austrian architect Konrad Frey (b.1934). Hunt had met Frey in Kingston-upon-Thames – Frey had worked for Arups in London.

Hunt's quest for reusable industrially-produced component-based structures continued throughout the 1960s and 1970s, culminating in projects principally in association with Michael Hopkins – an architect who shared with Tony Hunt an interest

[Opposite] *Distant view of Newport High
School – a Modernist concrete structure
for architects Evans and Shalev.*

[Below] *The courtyard of Newport High
School – the competition-winning design
by architects Evans and Shalev.*

in industrial, furniture and yacht design. One such project was the development of
SSSALU (Short Span Structures in Aluminium) – an all-extruded aluminium structural
system, which as well as offering a variety of structural solutions, included sliding doors
and used yacht rigging components for structural cross bracing. The design team for
this project included architect Ian Ritchie as well as Michael and Patti Hopkins and
Tony Hunt. The extruded aluminium system was taken to production stage by Heavy
Duty Alloys (HDA) of Workington, Cumbria.

Norman Foster had acquired a small pocket of land at Well Walk in Hampstead, London NW3, where he intended to build his own High-Tech house. After the break-up of Team 4, Norman and Wendy Foster had set up their own practice in Hampstead, which ultimately became Foster Associates. For a short time, Polly Hunt worked for them, prior to becoming a textiles specialist. She recalls the 'hands on' experience of working from the Fosters' home. Tony Hunt's work on this Foster prototype house in Hampstead in 1978 showed a much higher level of refinement than was evident in the SSSALU system. By using steel inserts, he was able to avoid the jointing problems that dogged the SSSALU system and the Richard Horden house that utilised aluminium yacht mast extrusions.

Michael Hopkins often mentioned that an ideal use for the SSSALU system would be for bunk house or summer house construction. This gives a clue to a possible antecedent for the design: that is the elegant Moduli summer house project by Finnish architects Gullichsen and Pallasmaa from 1969. There is a striking similarity between both the concepts and the graphic imagery of both projects. Tony Hunt was unaware of any design precedent for SSSALU, but he does recall the engineering difficulties and shortcomings of the extruded aluminium column to beam jointing cleat. Another project – a combined house and studio for a Dutch computer specialist – had taken SSSALU to the feasibility study stage of design, but it was never built.

In discussions of design precedents, Michael Hopkins referred to experience he had of lightweight building design in Scandinavia where, due to the fragility of flora in shallow topsoils in rocky areas, ideas had emerged to promote the use of suspended ground floors and minimal localised foundation work in order to preserve the natural status of the site and avoid damage to the landscape. Buildings in this vision would be placed carefully into position (even lifted and positioned by helicopter). Such thoughts were in accordance with his concept of lightweight design and with ideas for instant building. A direct comparison can be made between this High-Tech concept and the 1958 construction of a ski lodge by the self-build architect Walter Segal, where all components were transported to site using horse and sledge.

By the end of the 1970s Michael Hopkins was the principal exponent of industrialised component-based architecture of this type. Foster and Rogers had both co-operated with Hunt in various designs for lightweight systems, but they had moved on to building methods more in tune with clients and funding organisations for city centre developments. Last in the line of pre-engineered, prefabricated component-generated buildings designed by Hunt and Hopkins was the Patera Building – a larger scale steel version of the SSSALU concept, but which differed in that it was made from steel and intended mainly for small workshop buildings and could be relocated or reconfigured as necessary. In the late 1970s LIH – an industrial group based in Stoke-on-Trent, Staffordshire – commissioned designs for an 'off-the-peg' re-locatable industrial building made from steel. They wanted to expand their interests in steel fabrication and they intended to sell the buildings as a product. The factory where the Patera buildings were made and where the first two were erected was in Victoria Road,

Fenton, Stoke-on-Trent, Staffordshire. The idea of the Patera project was to supply an off-the-peg industrial workshop. The buildings were standardised – 18 metres long by 12 metres wide – with an internal height of 3.85 metres throughout. They were fully finished in the factory, ready for bolting together at the desired location. Three men with a forklift truck could erect one in a matter of days. Each building needed a reinforced concrete raft slab as a base to which the structure was fixed using specially designed steel castings. All the buildings' services – power, telephone cabling, water, etc – were distributed within the depth of the building envelope. The structural frames were pin-jointed for ease of handling and assembly. At the centres of the spans of the frames were 'tension-only' links – special fittings able to respond to varying structural loads. Under normal conditions the structure acted as a three-pin arch. In other conditions such as wind up-lift, it acted as a rigid frame. This innovation meant that very slender lightweight CHS (circular hollow section) steel tubes could be used for the portal frame trusses. The cladding consisted of structural steel panels. The same panels were used for both the walls and the roof. A new pressing method was developed for their manufacture that kept their edges flat and corrugated them in the centre, which increased their strength. All the panels were mounted on the inside of the frames and this brought certain advantages. By keeping the structure outside the building it was protected from fire should one break out inside, thus meeting fire regulations without the need to encase the structure with bulky fire resistant materials.

The first two buildings were erected at the site adjacent to the factory and stayed in place for some two years. They were used as demonstration buildings – part of the marketing of the project. Sites where other buildings were erected include Barrow-in-Furness, Canary Wharf and the Royal Docks in London. Only one Patera building remains at the time of writing. One of the three Patera Buildings built in Barrow-in-Furness, Cumbria was dismantled and re-erected in Broadley Terrace in London, to house offices of the architect, Michael Hopkins & Partners. Others have made way for property developments. The Patera project belongs to a small group of off-the-peg buildings by leading architects and engineers – other examples include the Wichita House (1946) by R. Buckminster Fuller (now fully restored and in the Henry Ford Museum), Maison Tropique (1956) by Jean Prouvé (which was brought back from Africa in 2008 after 50 years and put on show at Yale University and other sites internationally), and the IBM Travelling Pavilion (1984) by Renzo Piano Building Workshop. Interest in these buildings remains strong, perhaps because of their high level of detail and their 'collectable' toy-like quality.

Both Richard Buckminster Fuller and architect Walter Gropius produced designs for motorcars in the 1930s. Buckminster Fuller's was called the Dymaxion Car, which he patented in 1933. The vehicle confirmed his belief in lightweight engineering for both buildings and vehicles. He saw technology as the means to solve societal problems such as housing, food production and transport, so naturally his Dymaxion automobile was lightweight, highly engineered and original in concept and design. Gropius, however, carried out what was much more of a styling exercise for the solidly

[Above] *The tension-only link enabled a lightweight three-pin arched structure, the unique feature of the Patera Building.*

[Right] *Corner shot of the Patera Building with its all-steel construction.*

and traditionally-built Alder Limousine in 1931. Walter Gropius had been Director of the Bauhaus between 1919 (when it was called the College for Visual Arts in Weimar) and 1928. Here, in the Bauhaus spirit, he became interested in much wider design applications such as fabrics, furniture, ceramics, appliances, interior design, painting, graphics, books etc. Gropius described the ideal Bauhaus product thus: 'To serve its purpose perfectly ... to be durable, inexpensive and beautiful.' 'Inexpensive' was one of the qualities later described in Richard Hamilton's letter summarising the attributes of Pop-Art. Between 1934 and 1937 Gropius worked in England for Modernist architect Maxwell Fry before moving to the USA for his appointment to a post at Harvard University in 1937.

Although a direct comparison with Walter Gropius and Bauhaus would be misleading, it was clear that Hunt's move to Coln Manor in 1976 sought to create a more relaxed and 'free spirited' design environment than was possible in the hectic London surroundings. Under the tutelage of Felix Samuely he had experienced the same influences early in his own career that he sought to impart to his staff. He maintained a close interest in most forms of design, particularly in that of furniture. He readily recalls the group of colleagues who were prepared to follow him to Gloucestershire and this must be seen as one of the most energetic period of his career. As far as Tony Hunt was concerned the 'inexpensive' label for successful design – as referred to in the comments of Walter Gropius and Richard Hamilton – took the form of lightness of structure. The Buckminster Fuller philosophy of avoiding waste in the use of materials dictated the lightest possible structural solution, and therefore a relatively 'inexpensive' outcome. Although there often is such, there need not be any trade-off between cost and design quality. Gropius' ideal product would be inexpensive and well-designed. Amongst Tony Hunt's staff working at Coln Manor at the time was Mark Whitby, who went on to found the structural engineering consultancy firm Whitbybird. Whitby worked as project engineer on the Patera Building. Mark Whitby was to continue a tradition of steel house design for friends and architects. He carried out the structural design for architect and writer John Winter's house at Happisburgh, Norfolk.

5 FRIENDS, ARCHITECTS AND THEIR HOUSES

Tony Hunt was in the unique position of having shared in the 1960s adventure with a group of architects and engineers receptive to his ideas, particularly those centred around the progressive Architectural Association and the Royal College of Art, as well as those colleagues he met at Samuely's and at Hancock Associates. He counted this group of designers amongst his friends at the time. He was friendly with many of the London-based architectural writers and critics including Theo Crosby, Reyner and Mary Banham, Peter Cook, Cedric Price, Patrick Hodgkinson and John Winter. Tony Hunt recalls an occasion when he shared a platform with Reyner Banham at a conference in Norwich. They stayed at the house of a mutual friend and enjoyed swimming and relaxing together on the day after the conference.

The first house Hunt worked on was for Margaret MacDonald (Lady Casson). The house was named 'Matterys'. Hunt undertook this as a result of his meeting with Margaret MacDonald whilst working for Hancock Associates on the Shell Centre.

Sailing provided a common interest to Hunt and many of his friends. It took Hunt to Cornwall where so much early activity occurred with Team 4. It is said that Hunt's erstwhile employer Tom Hancock taught Hunt his yachting skills in Poole Harbour in a clinker-built sailing yacht. He recalls at least two holidays spent in the Poole Harbour area at Goat Horn Point with Tom and Jan Hancock. Tony Hunt first met Richard Horden when Horden was working at Foster Associates. Horden left there to form his own practice, which he originally set up in his garage. Hunt and Horden worked together on a project in Stag Place, Westminster, London, SW1 which eventually went to another firm of architects. Also, they worked together on various lightweight structures, but they didn't build anything together except Horden's house in Poole. Learning yachting skills here resulted in Hunt's love of the extreme and volatile kind of structural challenges that masts, sails and rigging demand. Hunt shared this experience with Michael Hopkins, who also enjoyed sailing and, owning and sailing yachts in the Mediterranean.

Hunt recalls that Norman Foster was distinctly uneasy whist he was introduced to the thrills of sailing. His reaction became a source of amusement to Hunt, who was himself by then an accomplished yachtsman, prompting questions to be asked mimicking the style of Foster's American business speak, such as: 'How is the ongoing marine experience assimilation situation?' Norman Foster found flying, cycling and running much more to his liking. He became an accomplished glider and powered aircraft pilot.

Tony at the helm of a 39 foot Dufor yacht, sailing towards St Tropez, photographed by Julian Hunt.

Tony Hunt's general design abilities, (meaning as well as structural engineering design ability) were held in high esteem by Norman Foster. In fact it was Hunt who designed the first Foster Associates letterheads and calling cards. They set up together an association called 'Consortium', which included Foster, Hunt and quantity surveyors D Hanscomb and Partners to seek work. Another venture intended to find work was a group called 'Component Consultants' that Tony Hunt set up with Martin Francis and Roderick Moody at 33 St Peters Square, London. Hunt produced letterheads for both of these ventures.

Martin Francis – a friend throughout Hunt's career – had worked for Foster Associates and was responsible for the glazing design at the Willis Faber Dumas building in Ipswich. He went on to design other sophisticated glazed structures before founding his own yacht design company. Tony Hunt bought a half-share in a yacht called *Anubis* (named by Martin Francis's wife after the Egyptian Mau cat) with Martin Francis before selling that share and buying another 30 ft yacht called *Amoret*. He moored and sailed the yachts in the Mediterranean for several years in the area of Antibes until 1976 when the proposed move to Coln Manor was planned and all available capital might be required for the move.

In 1970 Martin Francis and Tony Hunt designed the rig for the Rolling Stones European tour. There is a photograph of Mick Jagger, taken from the backstage area by Tony at the Olympic stadium in Helsinki. This was in September of that year.

[Above] *Tony working from his desk at 33 St Peter's Square.*

[Right] *'Anubis', owned jointly by Tony Hunt and Martin Francis.*

[Below] *The 30 foot 'Amoret', Tony's second Mediterranean yacht.*

Other colleagues of Hunt's at Hancock Associates were John Toovey (who was to become the London Zoological Society architect and was to commission Hunt in the design for the Sobell Pavilions at London Zoo), and for a short time Michael Hopkins. Hunt recalls that Norman Foster sought his counsel regarding Hopkins' suitability as an associate in the newly-founded Foster Associates. Foster had already met Michael Hopkins through Michael's wife Patti, who Foster had tutored at the Architectural Association.

As such, Hunt was party to the aspirations of the said group of friends – these were clearly expressed in designs for their own homes, where they were prepared to be more ambitious and less constrained by clients' wishes or whims.

A distinction should be drawn at this point between individual houses and housing schemes or developments. The transition from component-based structures and prefabricated, industrialised building methods to domesticity was a difficult one. Hunt had already successfully designed structures for housing schemes in concrete, brickwork and timber structures, but steel or aluminium were a different matter. The High-Tech architects were always nervous of housing where they had no control over the interiors, or how their designs would be used or viewed. Householders are usually very conservative and will react against designs and surroundings that are unfamiliar.

Foster Associates designed Bean Hill (with AHA as engineers) – a housing development in Milton Keynes for Milton Keynes Development corporation (MKDC). This comprised a series of timber-framed bungalows clad with black painted industrial profiled sheeting, and white painted exposed steel-covered walkways adjoining the bungalows' flat roofs. The outcome was a disappointment for all concerned. The flat roofs were quickly replaced with incongruous pitches in a mock-Tudor style and the industrial cladding was never popular with the residents. There was a national brick shortage at the time of this development, which might go some way towards explaining the choice of materials. It was completed in 1975.

From 1970 to 1976 Derek Walker (b.1929) was the first Chief Architect of the Milton Keynes Development Corporation. Specialising in urban design, he was responsible for the commissioning of some 3,000 houses each year and for the Central Milton Keynes Shopping Centre of some 93,000 sq metres. He was also Head of Architecture at the Royal College of Art. Tony Hunt worked with Derek Walker on the Advance Factory Unit (AFU) prototype building in Milton Keynes, which became the MKDC's architects' office. It can be said that even with the largest single architectural undertaking of that time – the new town of Milton Keynes – Hunt was involved at the formative stages.

The 125 Park Road, London housing designed by architects Farrell Grimshaw was also unusual as a housing development at the time, using visually uncompromising profiled industrial cladding set between ribbons of aluminium-framed windows. Nick Grimshaw lived in one of the apartments for about six years. The 11-storey block of 41 assorted apartments completed in 1970 was financed by funds from the newly-formed Housing Corporation. The building was listed as Grade II in 2001. Tony Hunt tells of an occasion when he was lowered by cradle and hoist to a depth of some 25 metres down one of the pile bores to inspect the condition of bedrock. He explained that the drilling had wandered off vertical and he needed to assess whether the load-bearing capacity of the finished concrete pile would be as he had intended in the design. This project avoided the criticisms aimed at Bean Hill. One possible explanation is that due to the multi-storey nature of the building, occupiers never came into close contact with the outside of the building above ground level and therefore never had cause to object to the industrial cladding. Nevertheless, this project cemented personal relationships with Terry Farrell and, of course, Nicholas Grimshaw, with whom Hunt went on to work in the future.

The visually uncompromising profiled industrial cladding, set between ribbons of aluminium-framed windows, which characterise the 125 Park Road multi-storey housing development.

The first of the individually designed houses for friends was Creek Vean – a concrete structure embedded into the Cornish coastal landscape. Team 4 architects designed this house and built it for the parents of Su Rogers – Marcus and Rene Brumwell. Hunt worked with Richard and Su Rogers on designs for the Design Research Unit (DRU) – of which Marcus Brumwell was the head – on a rooftop extension at Aybrook Street London in 1972. Hunt's first wife Patricia also worked there before working for the Arts Council and eventually becoming an accomplished architectural photographer. Tony Hunt remembers that there were partners at the DRU named Misha Black (who was a graduate of RCA), Milner Grey, Ken Bayes and Alexander Gibson. Also working at DRU was John Howard – a friend of Hunt's who sadly died in a car accident. Creek Vean is a family house built into the coastal landscape of the Fal estuary in Cornwall. Design of the project brought together Team 4 architects, which at that time consisted of

The trapezoid block of Creek Vean, photographed from the creek below.

Richard and Su Rogers and Norman and Wendy Foster in their first important building completed in 1964 (Georgie Cheesman – Wendy Foster's sister and original partner in the Team 4 practice – had left the group) and Tony Hunt – Principal of the newly founded Anthony Hunt Associates. Su Rogers' parents Marcus and Rene Brumwell commissioned the design. Marcus Brumwell was an art collector, head of a London-based advertising agency, and founder of the Design Research Unit – an organisation set up to promote British industrial design. The Brumwells had become friendly with a group of artists and had collected works that included pieces by Henry Moore, Barbara Hepworth and Ben Nicholson. They sought a Cornish property in which to display their works. It is reported that the Brumwells sold a painting by Mondrian to finance the building of Creek Vean.

The house takes the form of two trapezoidal blocks. The first is a two-storey living block and the other is a longer single-storey bedroom block. The two blocks are

Internally, the Creak Vean House maintains the concrete block construction to create split levels.

connected by a distinctive corridor with a glazed roof. It is this corridor that doubles as an art gallery. The corridor roof has patent glazing held into aluminium lay bars by neoprene gaskets. This is the only indication of the architects' engagement with advanced building technology that was to become the hallmark of the High-Tech movement. Between the two blocks is a set of steps down the hillside leading to a boathouse on the edge of Pill Creek. The corridor link passes at right angles below the external flight of steps with a solid roof over the corridor at their intersection. The trapezoid shape of the two-storey living block fanning out towards the shore, affords panoramic views – particularly from the upper floor over Pill Creek towards the Fal Estuary. The ground floors are built into the hillside and the bedroom block has a roof covered by vegetation, which is allowed to encroach upon the flight of steps. The impression is given of a house carved into the landscape. Slate floors are used on the ground floor to add to this impression.

Tony Hunt was able to produce designs for exposed structural load-bearing concrete blockwork together with a reinforced concrete frame that allowed wide-span panoramic windows in the living room block to be inserted using frameless plate glass. The execution of this traditional construction required the extensive and laborious cutting on-site of concrete blocks to match the off-square shape of the plan form. Surely this was anathema to the architects, who later built their reputations on batch-produced industrially prefabricated component-based designs, destined to become central to the High-Tech Movement. Nevertheless, the exercise demonstrated Hunt's ingenuity and willingness to work to realise the architects' visions, whatever the medium. He provided a well-resolved yet unobtrusive structure for a free-form design. The result was a building that became listed to Grade II status in 1997 and has been since elevated to Grade II*.

Creek Vean –, built into the Cornish landscape as it is –, demonstrates a preference popular in the 1970's to break down harsh geometric lines by the use of overhanging vegetation. This preference stayed with Foster Associates up to the time of the Willis Faber Dumas building in Ipswich. It is reported that some design drawings showed potted weeping willow trees overhanging the roof parapet onto the black glass facade. Folklore would suggest an occasion upon which Norman Foster directed operations from the street below, whilst porters carried the potted trees back and forth to create the desired effect. As a house for himself Hunt liked the Grade II listed 26 St Peters Square, Hammersmith property, where he had six people working at his first office. There he was conveniently close to his associate in Component Consultants, Martin Francis, who lived at 33 St Peters Square.

Tony Hunt had formed a series of friendships with architects, clients, and others including artists, photographers and industrial designers. These relationships date back to his time at Samuely's and Hancock's offices. Central to this group of friends and architects were Georgina and David Wolton. David Wolton was a merchant businessman and Georgina and her sister Wendy Cheesman (who married Norman Foster) were both architects and founding partners (with Richard Rogers and Norman

[Left] *Externally, a flight of steps leading down to a boathouse and the waters of Pill Creek is positioned between the two accommodation blocks.*

[Below] *Creek Vean, built into the Cornish coastal landscape, viewed from the waters of Pill Creek. (Image courtesy of Michael West)*

Foster) of Team 4. However, Georgie Wolton left the partnership almost as soon as it was founded. She had been the only member of Team 4 who had been registered to work as an architect at the outset. Hunt worked with Team 4, but also worked closely with Georgie Wolton and remained friendly. In fact it was David Wolton to whom Hunt turned when financial help and advice was required in his early days of independent practice. Hunt engineered structures for Georgie Wolton-designed houses at Cliff Road Studios on Camden Square, at 34 Belsize Lane and at Field House at Crocknorth in Surrey, where he designed a Corten steel-framed house for her. Hunt tells us that when Field House later changed hands, the new owners had no need for the steel house, and so Georgie Wolton arranged for it to be dismantled and for many years kept the frame in storage.

An interesting aside is that architects Spence and Webster had offices at the afore-mentioned Cliff Road Studios, and it was here that Tony Hunt became friendly with Robin Spence. Webster and Spence later won a design competition with Hunt for the design of the Parliament Extension building (an earlier scheme to the present Portcullis House). However that project was never realised.

Richard and Su Rogers' first commission after the break-up of Team 4 was the Spender House – a home for Humphrey Spender. He was an architect, artist and photographer who had studied at the Architectural Association but who had elected to become a photographer. Spender is well known for his mid 20th Century photographic project that documents social conditions in Bolton, Lancashire. Between 1937 and

[Above] *Snow covered ground adds drama to the reflective lightweight Spender House building.*

[Opposite] *Field House (a Corten steel-framed house) photographed from the garden.*

1938, Spender took over 900 photographs of Bolton (using a 35 mm Leica camera) at the request of Tom Harrisson, one of the founders of the Mass-Observation project. Spender referred to these as his 'worktown' photographs. One of Spender's associates at the Mass-Observation project was artist Julian Trevelyan (1910–1988). He painted the famous work *Bolton Market* as part of his study. The Spender House at Utling near Maldon in Essex represented a complete change from Creek Vean, made from a subtly engineered single-storey steel frame designed by Tony Hunt, with bolt-on industrial cladding. Hunt was required to design flat unobtrusive portal frames for the building. Where there would normally be a gusset plate to increase the frame depth at the point of maximum bending moment (that being at the eaves position), the architects demanded that the slim lines of the frames should be maintained, and so Hunt devised a substantial thickening of the outside flanges to maintain stability. He cites this as an example of architects dictating to engineers how a structure should look, regardless of the structural logic. It also shows the willingness of Tony Hunt to respond positively to such requests. It shows a shared belief in their aspirations for lightweight engineering design which appear lightweight.

Richard and Su Rogers' own house at Wimbledon – contemporaneous to the Spender House – took the bolt-together concept further in the use of proprietary composite cladding panels to produce a house similar in character to the Spender House. Though it was deceptively similar to the Spender House, the Rogers's Wimbledon house was far more sophisticated in its use of neoprene gasket seals, composite cladding panels and construction to exacting tolerances. The Spender

[Above] *The Spender House at Utling, Essex – a forerunner to the similar but more technologically advanced Rogers' House. The courtyard view is broadly similar to the Rogers' courtyard; yellow painted steelwork is a recurring feature.*

[Right] *Polly Lyster the textiles specialist.*

[Below] *The Rogers' House in Wimbledon took the bolt-together concept further towards a totally dry construction method.*

[Opposite page] *Interior shot of the Rogers' Wimbledon home; reflective surfaces and wholly glazed walls were a continuing feature.*

House was clad with industrial sheeting materials fitted on site using traditional methods whereas the panels used for the cladding of the Rogers' Wimbledon house were a proprietary product intended for use in the refrigeration industry. In the same way that yacht-rigging components would be used on buildings later, their use in this way represented what became known as technology transfer from one industry to another.

Having left Richard Rogers, Su Rogers later married John Miller – himself an Architectural Association graduate and RCA lecturer (becoming a professor there from 1975–1985) and a founding partner of Colquhoun and Miller. Hunt first knew Miller when he was at Samuely's and John Miller worked at Lyons, Israel and Ellis – the architectural practice that worked so closely with Samuely. Having introduced them to each other in the first instance, Hunt had continued his working relationship with them and designed the structure for their house – Pillwood House at Feock, Cornwall (at Pill Wood on the opposite shore of Pill Creek to the Brumwells' Creek Vean House). Here a steel frame supporting reinforced concrete floors and GRP cladding demonstrated the trend towards the High-Tech kit of parts that would become Hunt's hallmark. Hunt refers to some shortcomings of this particular design as it was very cold in the winter and suffered from excessive solar heat gain in the summer. He puts this down to the due south orientation of its substantial glazed areas and a general lack of insulation. As a lightweight structural design, Pillwood House demonstrates Hunt's prowess as well as any. Visually, it combines a light structural framework with carefully proportioned solid panels. Incidentally, the very first job that Team 4 undertook was in fact also on the opposite side of Pill Creek to the Brumwells' Creek Vean House, in Pill Wood. Marcus Brumwell owned properties on both sides of the creek. Known as 'the Cockpit', 'the Hut' or even 'the Retreat', this was a glass-roofed shelter built in woodland not far from the shore. Even though it is a simple lean-to structure of short span, the attention to detail and the appointment of Hunt to engineer the structure, pointed the way towards the bilateral architect/engineering co-operation which was to become the norm for future projects.

[Above] *Pill Wood House with its structure designed by Tony Hunt for John and Su Miller in Cornwall.*

[Below] *'The Retreat' – a well engineered albeit tiny structure in the secluded setting of Pill Wood.*

[Opposite] *The extensive areas of glass incorporated into Pill Wood House.*

The Sobell Pavilions – the Ape and Monkey House at London Zoo – brought Hunt into contact with another of his former Hancock associates, John Toovey. By now (1973) Toovey had become architect to the London Zoological Society – a position he held for many years protecting their rich design heritage. Interestingly, Hunt used a Mero Structures system for the enclosure – a firm with whom he was to work with again some 25 years later for the Eden Project. With David Tasker, Hunt devised a spring loaded fixing bracket for the lower edges of the mesh caging affixed to the structural frame. This allowed the monkeys to fling themselves at the mesh without weakening the fixings. Instead of using tension springs (as perhaps a trampoline arrangement might – this type of spring would run the risk of being over stretched and permanently distorted or broken), Hunt devised a method of using compression springs that could not be over-stretched. By use of a compression spring sleeved around a collared pin, the springs were protected from damage, despite the sudden and unpredictable dynamic forces to which they were subjected.

During the formative years of Anthony Hunt Associates, influences from the USA and Europe continued to impact upon Hunt's design philosophy. One such significant occasion was the unveiling of the Expo '67 exhibition event in Montreal, Canada. A geodesic dome designed by Richard Buckminster Fuller – the USA pavilion – confirmed the practicality and potential of such structures. Also at the exhibition was the West Germany pavilion: a tent-type glazed

Hunt reports that he has visited the site in Montreal recently (after he retired), and although the acrylic glazing to the geodesic dome had been destroyed by fire and removed completely, the structure itself has been renovated and well-conserved. The wonderful structure is also on view during aerial coverage of the Canadian F1 Grand Prix, with its waterside track in Montreal.

[Opposite page] *A semi-distant view of the Monkey Enclosure, built using Mero structural systems.*

[Right] *Steel mesh connected to the Mero structural framework by the specially designed sprung connectors.*

[Below] *Detail of mero structure and panel jointing for the Eden Project biodomes.*

structure by Frei Otto – a forerunner to the roof of the Munich Olympic Stadium built for the 1972 Olympic Games. Otto was working closely with Mero Structures at about this time.

In 1977 Hunt visited Paolo Soleri at the Arcosanti project in Arizona. He was impressed with the earth-formed shuttering technique developed there for the casting of mass concrete dome structures. He was also impressed by the techniques used for the earth formwork of flat facing panels, which were removed from their moulds and lifted into position. In the early 1970s Soleri had produced images of space colonies such as 'Asteromo' – a design for a space city 2,600 metres long with an internal surface of 466 acres supporting a population of 70,000. Such projects with origins in science fiction had also appealed to Buckminster Fuller, who in the 1960s envisaged

[Above and left] *Buckminster Fuller's 1967 geodesic dome conserved for future generations in Montreal, Canada.*

[Below] *The van den Bossche House in Fluy for Ursula Colahan by architect Ian Ritchie.*

[Opposite page] *Postcard of Coln Manor, Aerial view of the Manor House at Coln St Aldwyn, part of which was used as AHA offices.*

'Cloud Structures'. These were lighter-than-air globes 16 metres in diameter that were intended to float above mountain tops containing self sustaining communities.

Hunt designed a structure for the van den Bossche House for Ian Ritchie in Fluy, France. The house –built in 1976 for the parents of Ritchie's girlfriend Jocelyne – was a lightweight steel-framed construction with glazed walls and a traditional timber flat-roof. This combination of lightweight steel structure with a fully glazed perimeter but with a low-tech roof construction of timber joists and plywood decking reappeared in the Abbey Hill Golf Clubhouse in Milton Keynes for architects Michael Hopkins and Partners in 1982.

Tony Hunt's lifelong passion for industrial design and its relevance to structures prompted a search for offices away from London, where the creative design process could be carried out in relaxed surroundings. Designs could be tested and evaluated in the seclusion of studios and prototype or model-making workshops. In 1976 he bought a large manor house in the village of Coln St Aldwyn near Cirencester, Gloucestershire. He persuaded some of his London office staff to move there and also remarried Patricia. The move enabled AHA to set up one of the earliest computer systems for structural analysis. The system is reported to have been cumbersome with reams of print being produced according to a number of variables. It became necessary to search out the relevant solutions by marking key indicators in the body of an enormous amount of printed data.

In the tranquil surroundings of Coln Manor, Tony Hunt designed and prototyped two ranges of studio furniture. Hunt called his furniture the 'Alco Range', prototyped for him by Zeev Aram and built from 'Alucabond' – an ultra-light aluminium-based honeycombed sheet material intended for the building industry. This was a lightweight sandwich panel with aluminium facing. Hunt designed and prototyped a chair, table, chaise longue, and a workside trolley. He developed ways to form bends into the material. The resultant designs were never put into production but the prototypes remain in Tony Hunt's possession. Secondly, there was the 'O' range, which included a glass-top table, trolley with lacquered plywood top, and a director's chair fixed rather than folding. Hunt also developed a shelving system, which was simplicity itself – a folded aluminium sheet was formed into a 'Z' shape, giving a low down stand at the front of the shelf and a low raised backstop. Tony Hunt claims that Foster Associates made use of this simple design later on.

The decision to move to Gloucestershire was a protracted and complicated one. In 1974–1975 Tony and his wife had been looking for a country cottage in the Cotswolds. What they stumbled across was in fact an old manor house which had been Hatherop Castle School. There was an asking price of £55,000.00 but after a survey by David Allberry and some keen bargaining, a purchase price of £50,000.00 was agreed upon.

The part of the manor that the Hunts bought had been the Sixth Form Block. The use as schoolrooms lent itself quite well to an alternative use as studio offices, with what had been domestic science kitchens being converted into model-making and prototyping workshops. Hunt was able to preserve many of the original features of the manor house – exposed roof beams and masonry marks remained.

Hunt's idea was to move the offices of AHA lock, stock and barrel from London to Gloucestershire. The move prompted some lively debate and some reservations voiced by the staff in London. So a poll was arranged, giving each member of staff the chance to give their opinion. Hunt's own hunches as to those who might move and those who might stay were proved to be diametrically opposed to the actual outcome. Those he thought would want to move chose to stay and vice versa. In the end, 11 staff members moved to Coln Manor on April 1 1976. David Hemmings and John Austin opted to move to the Manor.

Alan Jones, one of AHA's longest serving Partners joined the firm while it was located at Coln Manor. The staff there expanded up to a total of some 45 people in the nine-year period there. It was in 1985 that the offices were moved again, this time into Cirencester itself. A property in Dyer Street was being developed into a range of ground-floor shops and first-floor apartments. An alteration to Planning Permission was negotiated, allowing office-use instead of residential at the upper level. The Gloucester House premises have been in use by AHA continuously since 1985. At the time of the sale of Coln Manor Tony Hunt married (for the third time) to Diana Collett. Hunt had met Diana whilst she was working for Michael Hopkins and Partners at their Broadley Terrace offices.

John Carter and Richard Clack had decided that Gloucestershire was not for them, and they remained in London. They continued to run the AHA London office from Bedford Street. In the end however, Clack and Carter parted company with AHA and set up together independently.

As the AHA no longer had a London address, it was decided in 1984 to share offices with service engineers Dale and Goldfinger in Blackfriars. This combined office was moved to Bay 8, Wharf Road, near Paddington Station – again sharing office space but this time with architect Dominique Michaelis. These offices were soon outgrown and a move to a ground-floor conversion in Chapel Street allowed an initial 30 or so staff to be accommodated. At this time with 60 or so in Cirencester, and 10 (including Steve Morley and Allan Bernau) in a newly-opened Sheffield office, AHA had become a widely respected and well-founded business.

Tony Hunt's career path entered a definitive period in 1976–7. Michael Hopkins left Foster Associates at this time and set up his own practice with his wife Patti. Hopkins has said that the failure to secure a transport interchange building in Hammersmith was one of the main reasons that he left Foster Associates. Tony Hunt was not involved in this project, but he felt the ramifications of the split between Foster and Hopkins. It was at this time that Tony Hunt found that instead of dealing with Foster and Hopkins as a single client, he had to follow each along divergent paths. He could no longer depend upon designs which up to this point had suited both of them. He needed to start afresh in order to cultivate good relations with both practices. He was helped to this end by a number of Foster's staff – such as Ian Ritchie and Mark Sutcliffe, who left Foster to work with Hopkins, if only for a short time.

From the many projects in which Hunt was involved, the most ambitious design of this era was the Hopkins House in Hampstead, London. Here, Tony Hunt produced an ultra lightweight steel and glass design. Unlike its design predecessors – IBM Cosham or Universal Oil Products, Tadworth – the Hopkins House was two-storey with the street entrance at the upper level. Completed in 1976, the Hopkins House was constructed as the home and studio of architects Michael (now Sir Michael) and Patti Hopkins. It has become a symbol that represents the design philosophy of a generation of engineers and architects that emerged in the London of the 1960s. Music, art and fashion produced an optimistic, technological and spontaneous culture, of which architecture was a part.

Built out of steel and glass, and engineered by Tony Hunt to the very limits of these materials, the structure appears slender and unobtrusive. The two-storey framework is a simple 2 metres by 4 metres grid of square hollow-section columns and lattice beams, these welded from hot rolled angles and concertina rods. The side walls are braced by double-skinned profiled steel cladding panels. Front and back walls are glazed with full-storey height-sliding toughened glass panels. From street level, there is a lightweight pedestrian bridge to the upper storey – the studio part of the house. The living areas are accommodated in the lower, garden level. The lightness of the design

The Hopkins House viewed from Downshire Hill, Hampstead showing the ultra-lightweight steel and glass structure. The pedestrian bridge leads to the first floor studio with living accommodation located on the lower floor. (Image courtesy of Steve Cadman)

prompted the famous American architect/engineer Richard Buckminster Fuller (on an occasion that he visited the Hopkins House) to ask his much-practised trademark question: 'How much does this building weigh?' During discussions about the design of the house, Patti Hopkins often used the expression 'wasteful', meaning that oversized components were to be discouraged and were aesthetically unacceptable. Efficient use of materials would produce its own engineering validity, and then its logical use would provide its own aesthetic. This was a manifestation of Buckminster Fuller's creed of lightweight design. It offered a way to make the best possible use of the earth's scarce resources.

There are unintended similarities in the building of the Hopkins House and Walter Segal's self-build method. Although the materials are different, Hopkins used steel and glass whereas Segal used timber. The self-build and low cost elements are common between the two. Hopkins has described the steel fabrication drawings as being his own, sent directly to a small blacksmith/fabricator, much in the way of Segal. Of course the philosophy behind the two could not have been more different, with Hopkins at that time demanding the crispest and most slender lines imaginable, and Segal seeking the most efficient methods of jointing standard materials with an inevitable trade-off between this and the quality and appearance of the finished buildings. Walter Segal was reputed to be an excellent speaker, enjoying group discussions, symposia and lecturing on his own subject. In this respect there are similarities with Richard Buckminster Fuller.

Michael Hopkins operated his house in much the same way as he might sail a yacht: constantly 'trimming' the performance of the controls by encouraging airflow and ventilation from the sliding glass panels to avoid heat gain, and by adjustment of the blinds to optimise daylighting, to reduce solar heat gain, or to provide privacy if required. By night the house could be made to appear light, bright and welcoming, or conversely quiet, restful and subdued. The low thermal mass of the house means that it warms up quickly by solar heat gain. Cooling is effected by airflow from the shaded garden, providing the greater cooling effect in summer. In winter when there is less shade, solar heat gain during the daylight hours is welcomed.

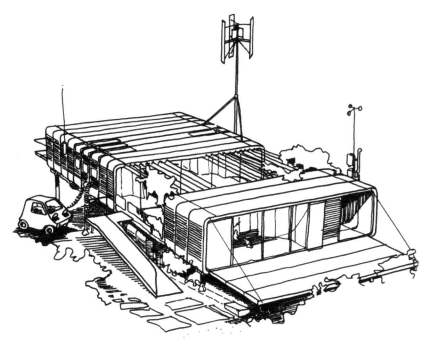

A similar double-height single-span aluminium structure was taken to prototype stage for Norman Foster's Hampstead home in 1978 – a project that was never realised other than by the mocking-up of the external structural elements. The design provided for a double-height double-cube single-storey unencumbered space with castellated lightweight aluminium pin-jointed portal frames positioned externally. The house had a flat roof and publicity images showed an array of communication devices located within the house, served by roof-mounted plant, aerials and a satellite dish.

The attraction that Hunt felt for this project owed much to the brief that Foster produced, whereby every element of the building from the structure down to furniture and fittings was to be designed afresh. This imagery – ahead of its time – foresaw what has now become commonplace. A furniture range named 'Nomos Table Series' designed by Norman Foster, had its origins here. With comparisons to space landing-craft 'legs' and 'feet', the furniture range became commercially available from Tecno, and was used mainly for offices. Tony Hunt still owns one of the first prototype pieces of the Nomos desk, which he still works from in the attic studio part of his house. It was built by a fabricator in the UK to exemplary quality standards, particularly the finishing of welded joints where two dissimilarly sized tubes meet. It is also worth mentioning that Tony Pritchard, who freelanced for Foster Associates (and who Hunt worked with again for Hampshire County architects), was a part of the design team for this project. Tony Hunt and Pritchard met up again when Pritchard worked on the glazing system for the McManus House, to a design by architect Peter Aldington.

The 'Zip-up' House – which should be referred to as the 'Du Pont' House, as they were sponsors of the design – was a conceptual design by Hunt and Richard Rogers for an off-the-peg house. It was the subject of a design competition sponsored by *The*

Daily Mail newspaper. This design, sadly never realised, took the house concept to its ultimate High-Tech 'plug-in' conclusion. A 'monocoque' shell construction from batch-produced structural panels would have been supported by jacking devices, avoiding the need for any wet trades during construction. The designs for the 'Zip-up' house clearly represented the current ideas for the house of the future. Imagery produced at the time was aspiring to efficient use of space, structure and materials using the philosophy of Buckminster Fuller and Jean Prouvé. Imagery of the 'Zip-up' House produced in 1971 provides us with conflicting glimpses into the future from the perspective of that time. The house itself envisages a mass-produced highly engineered off-the-peg design solution – a house that needs very little foundation work, that is expandable in size, and is factory made. In this image, a house is viewed as a mass-produced consumer item in just the same way as a car would be seen.

In the intervening four decades, the idea of a house being made in this way has all but vanished from view, but the image of a small electric town car being charged from a plug adjacent to the house has been realised. Advertising of electric cars such as the Mitsubishi-i-MiEV represents an almost uncanny manifestation of what was seen as the future in 1971. Technological development of off-the-peg house building has been left in the shadows. There is irony in this name, as 'plug-in' was always a term associated with the early High-Tech architectural movement, referring to the design possibilities of flexible-use accommodation and servicing modules. Now 'plug-in' has more to do with electrical devices.

The year 1971 also saw a project for the Aram standardised Hospital Module. This was envisaged as an off-the-peg hospital building for American clients. The architects were Piano and Rogers. The imagery reflected many of the Archigram-inspired High-Tech themes: open lattice steelwork, rooftop cranes for flexible assembly and reconfiguration, unitised accommodation, and minimal foundations. On more than one occasion, designs for such constructions by Hunt – and Rogers in particular – were eventually

Despite the unpopularity of this mechanised type of house from the mid 1980s onwards, similar ideas have re-emerged and are being promoted in the USA by architect Wes Jones. His Momo Redondo House project in 1999 reiterates many of the mechanistic images of Rogers and Hunt, but takes them further in concept by envisaging active mechanical operation and reconfiguration techniques instead of the manual re-configuration of Rogers's semi permanent designs.

compromised when the difficulties of satisfying funding organisations and building regulation authorities had to be faced.

Hunt's versatility and empathy was demonstrated in the design for the Eagle Rock House – an unusual house built for Ursula Colohan, Martin Francis's mother-in-law and designed by architect Ian Ritchie. Here, instead of a gridded system-built solution, lightweight suspension structures were used to provide a singular design tailored to fit a difficult undulating site.

With the propensity for individually designed houses from which architects could give clients a hint of their capacity for bespoke design, Tony Hunt co-operated

[Above] *Publicity images for Eagle Rock House for relatives of architect Ian Ritchie.*

[Left] *External view of Eagle Rock House from the garden, showing the house within its garden setting.*

[Below] *Wedgwood House – one of several steel and glass houses designed by Tony Hunt; this for architects Aldington, Craig and Collinge.*

with architect Peter Aldington – a founder of Aldington, Craig and Collinge. First of these was the Wedgwood House in Colchester, built in 1977. He was in touch again with Peter Aldington on the design of the McManus House at Princes Risborough, Buckinghamshire in 1982. On this project, Tony Hunt was not personally involved in the structural design, but he did visit the house during its construction as a guest of Tony Pritchard, who had designed the glazing system.

Tony Hunt was also acquainted with the artist Andrew Holmes, who had qualified as an architect through the Architectural Association, and who had worked for Richard Rogers in Rogers's small office after the break-up of Team 4. Although Holmes specialised in pencil drawings of mechanical and architectural subjects, he also produced front cover artwork for many Penguin Books, including *Rabbit is Rich* – one of John Updike's titles in his *Rabbit* series.

[Right] *The crisp lines of steel and glass: Wedgwood House.*

[Below] *Wedgwood House: Interior shot showing light-reflective surfaces complementary to steel and glass construction.*

6 THE HIGH POINT OF HIGH-TECH

Success in his support for the High-Tech architects brought Tony Hunt an array of commissions for a series of landmark projects. He became the automatic choice of engineer for many of the architects. Amongst these is Hunt's own favourite – of which he is justifiably proud – the now Grade I listed Willis Faber Dumas building in Ipswich, Suffolk. Foster Associates were the building's architects. Designed in the tradition of Owen Williams's *Daily Express* buildings – a glass box responding to street layout – Willis Faber had a curved irregular plan shape. Hunt's design solution was a concrete column and slab structure over three floors, with a stepped back steel-framed rooftop storey. Cantilevered floor slabs of up to 3 metres accommodated the irregular perimeter – the uppermost floor allowing suspension of the tinted reflective frameless glass curtain wall.

Willis Faber Dumas – a major insurance broking firm – decided to move its HQ out of London and selected Ipswich, Suffolk as a location with lower operating costs but with good transport links, placing it within easy reach of the capital. Their new offices were designed to accommodate 1300 people. Anthony Hunt Associates had worked with architects Foster Associates on earlier projects, but Willis Faber Dumas represented the first town-centre multi-storey prestigious building.

In concept the Willis Faber Dumas building owes much to the black-clad glass box buildings of the 1930s built for Beaverbrook Associated Newspapers in Fleet Street London, Manchester and Glasgow. These were designed by architect/engineer Sir Owen Williams and The Daily Express building in Manchester would provide the best example. The WFD building is planned to suit an existing irregular street layout on an island site. The building is three-storeys high, built to the back of pavement line. The roof is partly grass-covered and a steel-framed rooftop fourth-storey provides amenity accommodation for the building users.

[Below] *A night time shot showing the boldly designed cantilevered floor slabs with frameless glazing.*

Twin banks of escalators in the atrium below a glazed roof allow ease of movement throughout the complex. This can be seen as a forerunner of the concept used to dramatic effect in the Lloyds of London building. The striking reflective frameless curtain walling is the building's defining feature, mirroring the surrounding townscape in the day, but allowing a view of the interior and its activities at night – a technique used to promote the buzz and spectacle of night-time printing at *The Daily Express* several decades earlier. Deep piled foundations carry a square 14 metre by 14 metre grid of cylindrical concrete columns, which in turn support 700mm deep coffered concrete floor slabs. One metre diameter columns at ground-floor level reduce to 800mm diameter from the first to third-floor levels and reduce again to 600mm for the rooftop pavilion construction. 600mm diameter columns broadly at 7 metre centres form 'a necklace-like' secondary structure following the irregular plan form of the building, from which the floor slabs are cantilevered and tapered in thickness to the perimeter. The floor slabs reduce to 250mm at their edges. The ground-floor slab and swimming pool structure are supported between the pile caps. The roof-level pavilion restaurant has a lattice steel box truss-frame structure supported by the extended concrete columns – a familiar design, the product of continuing co-operation between Foster Associates and Anthony Hunt Associates. The Willis Faber Dumas building represented a turning point in the emergence of the High-Tech Movement, providing a step towards endearing the Movement to clients and funding organisations by the use of traditional concrete column and slab construction.

[Above] *The slenderness of the boldly designed concrete structure is clearly visible at night.*

[Opposite] *An internal view of the Willis Faber Dumas building's atrium, looking down from the grass-roofed restaurant level to the floors below.*

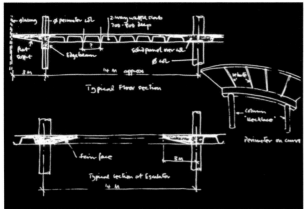

[Above] *Tony Hunt's sketches for the logically-generated structural grid to resolve an irregular plan shape.*

[Right] *The Willis Faber Dumas entrance area with a view up the range of escalators to the atrium's glazed roof.*

[Opposite page] *Ben Johnson produced a painting of the ground floor interior of the Willis Faber Dumas building.*

It is worth mentioning that Martin Francis – a lifelong friend and sailing companion of Hunt's – was working for Foster Associates and devised the curtain walling fixing methods for the Willis Faber Dumas building.

As the construction of the Willis Faber Dumas building was reaching its critical stages, reports reached Hunt of a possible structural failure. David Hemmings from AHA travelled to the site and identified the fabrication failure of a diagonal compression member within the box beams at the roof-level restaurant. The fault was reported to the fabricator Tube Workers, and under the supervision of Mark Sutcliffe from Foster Associates and David Hemmings from AHA, the problem was rectified.

Also for architects Foster Associates, Hunt engineered the Sainsbury Centre for Visual Arts, built on the University of East Anglia (UEA) campus in Norwich. The epitome of High-Tech design, this was a steel lattice-framed single-span supershed, with plate glass end walls and clad with aluminium panels. Bolted together and sealed with EPDM gaskets for both walls and roof (gaskets were formed into gutters for the roof), Hunt contributed to the advanced building technology which gave rise to the High-Tech label. On the UEA campus, the Sainsbury Centre for Visual Arts is a near neighbour to other parallel symbols of Modernism: the teaching hall, raised walkways, a central square, and the ziggurat-form accommodation blocks of Norfolk and Suffolk Terraces. The core of the UEA campus was designed by architect Denys Lasdun. He envisaged a campus given over to pedestrians, where any part of the campus was within a five-minute walk. Whereas Lasdun's uncompromising concrete designs are derived from the European Modern Movement from his early career with Berthold Lubetkin and the Tecton group, they are juxtaposed with a High-Tech design with its own transformed Modernism, derived from the American influences of prefabrication, dry construction, flexibility and adaptability that informed Foster Associates designs from Norman Foster's time as a student at the Architectural Association and at Yale University School of Art and Architecture.

Sir Robert and Lady Sainsbury donated their art collection to UEA and had sufficient confidence in the designers to entrust their patronage to an ambitious steel and glass structure clad with bolt-on aluminium panels. The Sainsbury Centre for Visual Arts comprised two exhibition galleries and a School of Fine Arts, together with the necessary ancillary accommodation. Their brief for the building was exact, based upon their experience of having visited many galleries throughout the USA and Europe. Panels specially designed and manufactured for this project alone, are standardised to a size of 1.8 metres by 1.2 metres and have EPDM (ethylene propylene diene monomer) gasket seals, which were welded on site into a net covering the whole building. The panel jointing gaskets doubled as guttering, discharging rainwater from the roof to the base of the building. The aluminium panels were vacuum formed to

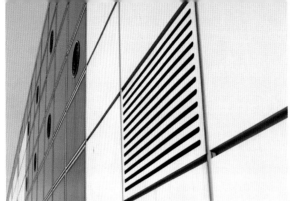

Standardised panels are used for both the walls and roof of the Sainsbury Centre for Visual Arts. (Images courtesy of Martin Pearce)

provide strengthening ribs into flat sheets as a part of the fabrication process. End walls are constructed from 7.3 metre-high glass panels, jointed with silicon mastic. This technique pioneered the way for frameless glass walling on such a large scale. The internal dimensions are generated from 36 bays of 3.6 metres (the glass walls are set in by a bay at each end). The internal height is 7.2 metres and the internal clear span is 28.8 metres. The building is logically planned from inside to out and is formally proportioned in a length/width/height ratio of 16:4:1. All plant and services are contained within the depth of the structural zone giving uninterrupted planar surfaces to both the inside and outside. Inside walls and ceilings are formed by adjustable motorised aluminium louvres. Supplementary artificial lighting is sourced from within the ceiling void – also controlled by the automated louvers – giving constant uniformity of lighting levels to the gallery spaces. There is a spectacular access to the building via a raised walkway leading to the upper level within. A very slender spiral staircase down to the galleries completes the dramatic entrance. The sequence is reminiscent of the escalator access to the Dome of Discovery for the Festival of Britain centrepiece building (designed by engineer Ralph Freeman and architect Ralph Tubbs).

The single-span structure of the Sainsbury Centre is built up from frames comprised of three-legged tower columns made from welded CHS steel tubes (two legs aligned with the outer facade and a single leg positioned on the line of the inner louvered walls). These lattice columns are cross-braced to match the 1.2 metre panel heights and spaced to align with every other 1.8 metre bays of panelling – that is at 3.6 metre centres. The simply supported cambered roof beams are also made from welded lattice steel prismatic trusses. These are supported by a carefully positioned platform plate at one end and a pin joint at the other. The platform plate and pin joint are carefully positioned to allow adjustable connections during the assembly process and importantly to facilitate curved cladding panels at the eaves. Tony Hunt has said that this assembly allowed manufacturing and site tolerances to be accommodated and yet allowed the imagery of a seamless portal frame of uniform dimension. To this end, the structure has been adapted to suit an architectural vision. The Sainsbury Centre for the Visual Arts is an outstanding example of the High-Tech vision of engineering and architecture and one of the few to have used a bolt-together interchangeable panelling system.

Tony Hunt recalls an episode concerning the staircase intended to create a dramatic entrance into the gallery areas from the raised walkway. The initial designs called for a free-form stepped ramp, parabolic in shape, but the structural demands of such a shape proved to be insurmountable. The outcome was a spiral staircase designed late in the day, by which time the building shell was substantially complete. The fabricated steel staircase was introduced by 'winding' it in through the lower-level

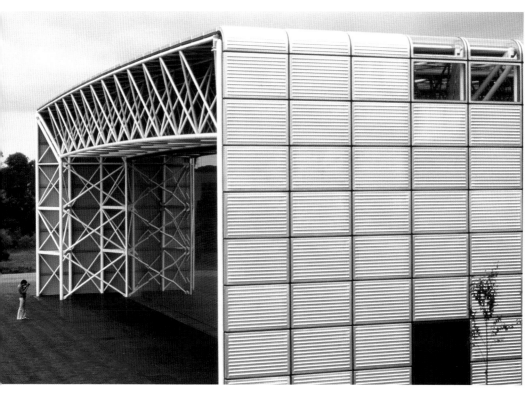

The panelled side to the Sainsbury Centre for Visual Arts, showing door and window panel types as well as the standard solid panel. The 28.8 metre clear span is designed and fabricated to include a modest camber to the prismatic roof trusses to aid rain water run-off.

entrance doors, 'like a corkscrew action' as Hunt describes it. After completion of the galleries and after the art works had been put on display, it was noticed that the spine of the staircase – a CHS steel tube – was bending to an unacceptable degree. The art work was temporarily protected whilst roof panels were removed to allow stiffening ribs to be lowered in through the roof opening and welded into position. This operation required the use of a mobile crane with exceptional reach.

Hunt was able to continue to respond to demands for the engineering of industrial architecture in a similar theme to Team 4's Reliance Controls and Foster Associates' Modern Art Glass building and work for the Fred Olsen shipping line – a passenger terminal and headquarters building at Millwall, London (also with Foster Associates). For the Greene King Brewery building at Bury St Edmunds, Suffolk he worked in association with Michael Hopkins architects. Built above the flood plain of the river Linett, a concrete column and slab suspended ground-floor was constructed below the lightweight lattice steel beam and column superstructure. The envelope was made up of profiled steel cladding and glazed industrial doors. This building – though less technologically advanced than others of this period – was carefully planned and detailed, winning design awards for architecture and structural engineering. Covering a gross area of 3250 square metres, the racking plant (racking is the process to put draft beer into casks) is a highly-serviced process-plant building, where empty casks are brought by road vehicle to one end of the building to be cleaned out and filled and then dispatched at the other end, again onto road vehicles. The floor level was set at tailboard height above the surrounding roadways, enabling efficient vehicle loading and unloading. The simple rectangular plan form represents this process. Beer is stored in tanks filled by a pipeline running from the brewery itself. The structural solution is twofold. The raised ground-floor is suspended above the flood plain by a 2.5 metre by 2.5 metre grid of circular concrete piles cum stub columns. The in-situ cast floor slab is contoured to provide drainage channels for the run-off of spillage and for the purpose

[Opposite] *The single-span structure of the Sainsbury Centre showing the carefully positioned platform plates at the eaves.*

[Below] *An interior view of the Sainsbury Centre for Visual Arts showing the troublesome staircase.*

of floor cleaning. It also provides fixed bases for the plant and process machinery. The superstructure comprises circular steel hollow section 6 metre high columns on a grid (compatible to the concrete columns below) of 10 metres by 15 metres. The columns support trussed beams, also made from circular hollow section welded steel of similar structural depth for both 10 metre and 15 metre spans. Cantilevered extensions of 1.2 metres provide cover over pedestrian walkways along each side of the building, with a 5 metre cantilevered continuation of the flat roof, covering loading platforms at the front and back. Symmetrically positioned staircases protected from vehicle movements provide pedestrian access to the walkways at each corner of the rectangular building. Double skin profiled steel cladding uses the profiling to span horizontally up to 10 metres. Extruded aluminium edge supports are positioned on the 10 metre gridlines externally, but at closer centres internally. Front and back elevations are comprised of 'up and over' industrial doors, suitably stiffened to operate over the required 7.5 metre width.

The offices of Michael Hopkins Architects at 27 Broadley Terrace, London, NW1 are housed in buildings constructed from components adapted from those developed for the Patera Building – an industrialised prefabricated building made entirely from steel and glass and envisaged as an off-the-peg workshop. As mentioned in Chapter 4, Anthony Hunt Associates were engineers for the Patera Building and they were able to adapt the structure to suit an entirely different building. The main building at Broadley Terrace (the Number 1 Building) has a floor area of 324 square metres, 18 metre square in plan. Its internal height of 5.05 metres is sufficient to allow free-standing mezzanine floors to be introduced. Similar in design to the Patera Building, the envelope is made from 3.6 metre

by 1.2 metre pre-finished double-skin insulated pressed steel panels, used both for walls and roof. These are carried by 100mm x 50mm RHS (rectangular hollow section) purlins connected by special fasteners through to the main structure at 1.2 metre centres. Unlike the original, Building Number 1 has 1.2 metre deep lattice roof trusses supported by cantilevered columns located at 3.6 metre centres externally along the sides of the building. CHS line bracing stabilises the roof trusses also at 3.6 metre centres, with stainless steel rods (with adapted yacht rigging connectors) providing cross bracing. Foundations and cantilevers for the columns were calculated to avoid the need for any cross bracing for the columns, relying only on the diaphragm action

[Above] *Docklands landscape reflected in the frameless glazing of the Fred Olsen building.*

[Below] *The stark industrial and mechanistic end elevation of the Passenger Terminal.*

[Opposite] *The Fred Olsen Passenger Terminal viewed from alongside it.*

of the panelling for stability. The choice of a totally different structural system – column and trussed beams instead of the three-pin arched hybrid structure developed for the Patera Building – posed a challenge for both the engineers Anthony Hunt Associates and the manufacturers. Although appearing to be flat, the roof trusses are fabricated with a camber (with a rise of 90mm) to the mid span for rainwater run-off (the three-pin arch of the Patera Building allowed a slight pitch to the mid point to allow rainwater run-off). At Broadley Terrace, movement due to deflections from imposed loads (such as snow loads) is accommodated in movement joints at the head of the glazed back and front (street frontage) elevations.

The introduction behind Building Number 1 of a standard off-the-peg Patera Building (216 square metres, 18 metres by 12 metres and 3.85 metres high) relocated from an earlier location in Barrow-in-Furness confirms the validity of the Patera concept – i.e. relocatable and ease and speed of installation (even with the restricted access available at Broadley Terrace). This Patera Building was reconfigured to include glazed panels and entrance doors in the side, and provided an annexe to the main office. The engineering for the relocation of this standard building was carried out by Mark Whitby of Whitbybird, who had been closely involved with the design of the Patera Building whilst working for Anthony Hunt Associates in the early 1980s at Coln Manor. Mark Whitby's definitive contribution to the design was the 'tension-only' link at the mid-span of the externally positioned trusses.

The Broadley Terrace development is completed by a fabric roofed walkway providing a main street entrance and a covered link to each building. This layout reflects that of the original brick building on the site. The fabric roof is extended to provide external shading to the glazed centre bay roof of building Number 1. This walkway effectively draws together the disparate nature of two similarly prefabricated buildings to produce a well-planned city studio representative of Michael Hopkins Architects' technology driven designs.

In 1983–1984 Hunt had decided to return to London after the breakdown of his second marriage to Pat Hunt. The Coln Manor property was to be sold, and so when Hunt moved back to London he became the first tenant for the guest rooms at Robin Spence and Robin Webster's studio house in Belsize Park, North London. The house was completed in 1981 but in a twist of fate, the engineering for this project had been carried out not by Tony Hunt himself, but by his former partners at AHA – John Carter and Richard Clack.

Tony Hunt recalls an incident whilst he was living as a tenant with Robin Spence: there were regular meetings about once a month, when members of an informal group would give a short presentation. The group comprised of like-minded architects

Street view of 27 Broadley Terrace, the Patera-constructed offices of
Michael Hopkins and Partners. Hopkins' Porche 914 is in shot.

including Robin Spence, Richard Horden, Tony Hunt, Martin Francis and Michael Hopkins amongst others. Robin Spence organised these gatherings and on some of the occasions Tony Hunt was required to vacate his temporary lodgings so that the talks could go ahead in his part of the house. Hunt recalls Patti Hopkins' look of surprise

Interior view of 27 Broadley Terrace's five metre clear height glazing to front and back.

when Hunt arrived in attendance of one such talk at the Hopkins House accompanied on his arm by Diana – their accounts manager. Martin Francis was giving the talk on this occasion and for reasons that Tony Hunt cannot remember, the venue had been changed from Spence's house to the Hopkins House in Downshire Hill.

After a short time as tenant to Robin Spence, Tony Hunt moved again to the Studio House, 14 St Johns Wood Road, close to Lords Cricket Ground. In a serendipitous moment, he had acquired the studio from another architect friend – Mike Davies – who was a Partner at Richard Rogers Partnership. Hunt just happened to be driving past the property one day when Mike Davies was getting ready to move out. The deal for Tony Hunt to buy the property was done there and then on the pavement. Mike Davies had studied at the Architectural Association and had joined Piano and Rogers shortly after the practice had been awarded the Pompidou Centre project.

Hunt took the opportunity to work on two projects in the 1980s using lightweight structures of wood instead of steel or aluminium. At the Lindisfarne Chapel at Crestone, Colorado, USA, he engineered a laminated timber structure where the interlaced shaped beams were formed and bonded on site. The dome there was constructed without any ferrous material. The architect for this project was Keith Critchlow – professor at the RCA and a specialist in Islamic art and architecture. Hunt had come across Critchlow during his time lecturing at RCA, but never had any substantive dealings with him until Crestone. The second project was a specialist construction for the Halley Bay in the Antarctic – a tubular stressed skin of multilayered plywood. This was designed during Hunt's time at Coln Manor. The architect for the project was Angus Jamieson. Mark Whitby of the AHA staff had design input into this project. The British Antarctic Survey has maintained a base in the Antarctic for several decades. It is a recurring feature of High-Tech architecture and engineering to match design excellence with extremes in environment. In hostile locations such as the Antarctic, high-performance components are essential as is lightweight construction for easy transportation of components. The twin themes of lightweight structure and high-performance components were central to the High-Tech Movement, but only occasionally were the environmental demands an appropriate match for the concept. In the extremes of environment encountered during aircraft flight, in the Antarctic, in high Alpine weather conditions, or in ocean-going yachts, there can be no acceptance of inferior component design or of over-heavy structure. This willingness to embrace the design solutions for extremes and to transpose their essence to the everyday informed the High-Tech Movement in architecture and engineering. These two projects reflected Hunt's early expertise in timber engineering gained at Hancock's when he engineered complicated curves for the laminated roof beams of the Carmel College synagogue.

[Above] *Secondary framework in position awaiting cladding.*

[Right] *Horizontally-laid boarding brace the arched lattice structure.*

[Above] *Practical application of the complex geometry by volunteer labour.*

[Right] *Primary ribs shaped around framed centring.*

[Above] *Interior showing rhythmical patterning of the compound curves of the timber structure.*

[Left] *Architect's model demonstrating the build-up of geometric structure and cladding.*

[Below] *The multi-layered plywood skin of the Halley Bay British Antarctic Survey building during construction.*

[Below left] *The Halley Bay building in use, partly covered by the drifting snow and ice.*

In a more traditional engineering role, around this time Hunt engineered a trussed timber roof for the Newlands Primary School at Yateley, Hampshire, under the enlightened patronage of Colin Stansfield-Smith – Hampshire County Architect. Stansfield-Smith made a reputation for himself and for Hampshire County Council by acting as patron as well as running his own very successful county architects department.

Tony Hunt worked with architect Ted Cullinan (b.1931) on a headquarters building for Ready Mixed Concrete (RMC). The design – no doubt prompted by the clients' business – comprised of a coffered concrete structure surrounded by an exposed steel perimeter structure. Built between 1986 and 1990 in the Surrey landscape at Thorpe Park, RMC addresses energy conservation issues by providing a grass roof over naturally ventilated offices. Another of Hunt's long-standing associates – Max Fordham – engineered services. Ted Cullinan was educated at Cambridge, the Architectural Association and at the University of California in Berkley. He worked for a time for Denys Lasdun. One of Cullinan's first jobs was a house for his uncle called the Horder House (built in 1960) in Camden. Here, Cullinan developed a low cost single-storey timber structure, similar to the work of Walter Segal. In Cullinan's case, precedents for this type of lightweight structure may have come from the Californian Case Study houses that he would have seen whilst studying there, whereas Segal's had European influences. Foundations for the house were inexpensive and minimal: lengths of asbestos cement tubes set vertically into the ground and filled with concrete. Directly onto these stub piles a suspended ground floor was placed, and onto the projecting floor beams the superstructure was built.

As mentioned above, by 1985 Hunt's second marriage to Patricia had ended and he decided to move out from Coln Manor back to London. Later that year he was married (for a third time) to Diana Collett. With Diana, Tony Hunt once again sought the less-frantic surroundings of Gloucestershire. He moved to 5 Cecilly Hill in Cirencester and again to Overley House. The next house they moved to was an Arts and Crafts style country house that Hunt liked, called Warrens Gorse House. It was designed by architect Norman Jewson (1884–1975). Jewson worked from his offices at Sapperton in the Cotswolds. Richard Horden and his wife Kathy were close friends of Tony Hunt and of Diana, Kathy being Diana's best friend. It was a tragedy when she died as a result of a riding accident in Hyde Park. Richard and Kathy's daughter Poppy Horden is Diana's Goddaughter. Tony and Diana spent some time with them on Green Island in Poole Harbour. Richard Horden and Hunt went to Lauterbrunnen to re-erect the Ski Haus. They directed that it should be transported up into the Alps, lifted there via helicopter. Hunt wrote an account of this for *Building Design* magazine.

Built in 1991–1992 the Ski Haus serves as a mobile alpine hut or 'hard tent'. Its lightweight structure weighs only 315kg. It is well-insulated using aviation materials and has self-sufficient energy systems powered by solar and wind generators. Since May 2004 the Ski Haus has been situated on the Swiss-Italian ridge close to Kleines Matterhorn. It is used as a shelter for climbers and skiers. There are several antecedents that the Ski Haus can be compared to. Firstly there are similarities with architect Walter Segal's first project – a ski lodge made from components transported by horse and sledge up the mountain, rather than by helicopter. Secondly, with the Ski Haus the imagery of Future Systems' High-Tech architecture is reflected in a building owing more to the aircraft industry than to the building industry, but in this instance the Ski Haus is a real structure and not an imagined one. And thirdly, we can see that such an extreme environment such as that found high in the Alps, calls for a High-Tech building solution. These extremes are not to be found in cities or urban areas, where a conservative approach to building would inevitably tone down a High-Tech solution.

[Above] *Ski Haus being transported from the heliport at Lauterbrunnen to the north ridge of the Schreckhorn in the Bernese Oberland. (Image courtesy of Horden Cherry Lee Architects)*

[Opposite] *Diana in relaxed pose, photographed by Tony in 2004.*

[Above] *Tony Hunt helping with the installation of Ski Haus. (Image courtesy of Horden Cherry Lee Architects)*

[Right] *Air Zermatt helicopter similar to that used for the installation of the Ski Haus. (Image courtesy of Horden Cherry Lee Architects)*

From 2001 Richard Horden has undertaken the role of Professor for Architecture and Product Design at The Technical University in Munich. Student projects were published from Munich taking his Ski Haus concept some stages further forward with a Peak Lab Research Station. Innovative and daring structures and construction methods reflected many of the British High-Tech architects' concepts that Horden had been party to earlier in his career.

With Diana, Hunt then bought the 100-acre Stancombe Farm in Gloucestershire. But alas, after only about four years together there, the marriage failed. Hunt reluctantly sold the farm and moved to his present studio cottage in Box in the familiar surroundings of the Cotswolds. Built in Cotswold stone down to the last detail of a curved staircase, Tony Hunt's studio cottage is filled with furniture, drawings, artwork and mementos representing his lifetime of work in design. On the ground floor is his kitchen living area with his son's refectory table, surrounded by model aircraft, an 'O' gauge wind-up railway locomotive, photographs of his yachts and many other keepsakes. There are some framed sketches by his hand, similar to those exhibited at the Chelsea Art Club in 2009.

Even to this day, Hunt is still designing – notably wall-mounted light fittings for the cottage, using industrial components. On the ground floor in what was a coach house is Tony Hunt's library. On the first floor, there is a living room containing a Saarinen 'tulip' table, an LC4 chaise-longue, Eames' chairs, Breuer chairs, Alvar Aalto stools, and on the book shelf, a model of a Reitveld chair. On the top floor, Tony Hunt has an art studio with his prototype Foster Nomos desk, wire-framed shelving and a Quad 44 405 and FM 4 sound-system of classic British design from the late 1970s or early 1980s. Of note is the Fred Scott (1942–2001) designed 'Supporto' task chair that accompanies the Nomos desk. This is a British cast aluminium chair, which can be compared to the much more famous Eames-designed cast aluminium chairs. Tony Hunt met Scott when they both lectured at the RCA furniture course – Hunt says that Scott's design is more comfortable for extended periods of desk-work than Eames'.

7   HUNT'S QUEST FOR TECHNOLOGICAL ADVANCE

In the 1960s and 1970s Hunt had been able to realise the designs of the High-Tech architects, giving credibility to their untested designs. The result of this continuing co-operation meant that Hunt was called upon to produce innovative designs for each of the High-Tech architects. Thus, the responsibility for technological innovation was thrust upon him. He became the engineer of choice to the High-Tech architects and was expected to produce ever more ambitious design solutions. Under the Buckminster Fuller creed of lightweight structures, Hunt achieved success in a progression of technological advances, with the Buckminster Fuller philosophy to reduce weight and make more slender structures.

It is worth mentioning here that a distinction should be made between different types of technological advance. Technology transfer – where techniques or processes from one industry are introduced to another – is the first example. Such might be the Rogers 'Zip-up' House, where proprietary composite cladding panels designed for the refrigeration industry were used, or yacht rigging components used in tension for structural cross bracing. A second area of technology transfer is identifiable where products from the building industry are used in an unconventional or novel way. This would be an architectural choice without any structural requirement for use in this way. Hunt was familiar with the use of profiled steel cladding to span horizontally instead of vertically. Greene King Brewery and Hopkins House both use this technique. A 'fad' or 'fashion' might be a better description of this particular trend. Foster, Rogers, Farrell Grimshaw and Hopkins all used profiled sheeting horizontally at one time or another. Hopkins held onto the fashion longer than the others. In this particular technological development, the manufacturers responded by producing a profile of a more suitable shape to be used in this way. Technology transfer in the case of profiled sheet cladding is demonstrated when used as the material for houses and other small-scale buildings instead of on larger industrial buildings for which it was intended. In other words, inexpensive, readily available, mass produced materials were used out of context. The third example of technological innovation is where manufacturing technology is harnessed with a definite structural or architectural image in mind. Tony Hunt's quest for technological advance was most evident in his dealings firstly with architects Foster Associates, Michael Hopkins and Partners and then with Nicholas Grimshaw and Partners. Vacuum-forming aluminium panels for the Sainsbury Centre for Visual

Arts building at UEA, Norwich was an early example of this process. Advances gave architects Foster Associates (which at that time included Michael Hopkins) an architectural solution for this one project. Mass production of panels to standard size allowed variation of panel type, interchangeability, removal and replacement of the standard panel. The structural requirements for the panels themselves were modest, but nevertheless required Hunt's input.

Tony Hunt was never more at home than when he was devising and designing innovative structural components. His friendship with Martin Francis arose because of their shared love of technological advances both in the building industry (Francis was a part of a team originally at Foster Associates, which devised advanced technological solutions) and in the boat building industry. The prototype house in Hampstead – designed in 1978 with and for Norman Foster – provided Hunt with a further opportunity to design innovative structural components. Hunt had devised an aluminium pin-jointed column and beam-frame structure for the single-span. Aluminium was a natural choice due to its superior finishing, anti-corrosion, weathering and workable qualities when compared to steel. However, with an inferior strength-to-weight ratio than steel, Hunt devised methods of stiffening the aluminium with tubular steel cores where jointing was required. He made possible primary and secondary structural connections with a range of standardised joint designs. With this technological advance, he was able to maintain slender elegant proportions to the structure without resorting to additional bulk. Although the Foster house was never built, Foster Associates developed the technique of wrapping steel structures in aluminium in later projects.

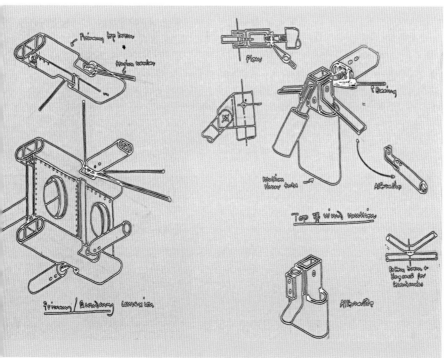

Hunt's sketches of the Foster prototype house, showing the aluminium structure stiffened with steel inserts.

Further innovation arose from the SSSALU aluminium building system, which required Hunt (with Hopkins) to calculate the structural properties of entirely new aluminium extrusions, their jointing techniques and bracket or cleat design. Although a more thoroughbred design in its use of extruded aluminium to solve both structural and weathering problems, SSSALU failed to match the elegance of the Foster prototype house structure. On designs for the Patera Building, Hunt (again with Hopkins) was able to push the bounds of technology in two areas. First, the steel structural cladding panels were specially pressed to give corrugations capable of spanning between purlins whilst maintaining a flat square edge for weather sealing, and to give stability to provide a diaphragm action between frames. This avoided the need for a secondary external bracing to the structure. The pressing process called for the use of pre-galvanised steel sheet of EDD (extra deep drawing) quality. Zintec or EDD pre-galvanised sheet were the only pre-coated steel that was found to press successfully. The press – at 1800 tonnes with a cushion weight of 50 tonnes – was of the type used in car body panel pressing. In order to keep the edges and flat areas of the panels flat, an additional pair of ribs were included each side of the main corrugations. This rib, which was formed as the cushioning effect of the press took hold, clamped the sheet into a fixed position as the main weight of the press acted. The 3.6 metre long panels were pressed in three such strikes in order to reduce the size and cost of the tooling itself, leaving the strengthened sheet with flat edges and surfaces ready for further processing. The same press form was used for both inside and outside surfaces of the double-skinned structural panel. Secondly on the Patera Building, the cast steel components for jointing the lattice steel trusses provided advantages in accuracy of fabrication – castings could be introduced into assembly jigs guaranteeing accuracy in set-up prior to the welding operation. Jointing pin positions were replicated with machined dummy pins on the assembly jigs, ensuring that accurately drilled holes (which would accept the pins during construction) on each component were identical and their key contact points were precisely located. Also, particularly in the base plates (with provision for water run-off and positioned shoulders for the holding down bolts), castings provided technological advances that traditional fabrication could not have equalled economically.

Buckminster Fuller's influence on the High-Tech Movement was profound. It provided the participating architects with their creed: that design should be lightweight, and hence should not be wasteful in their use of materials, to be frugal and thus conserve Earth's scarce resources. In 1946 Buckminster Fuller patented the Dimaxion Map also known as the Fuller Map. By superimposing a geodesic grid pattern onto Earth's surface and unfolding the three-dimensional form, Fuller produced a series of two-dimensional maps which he claimed gave a much more accurate representation of Earth's surface. From this accurate map, he felt better able to evaluate the extent of Earth's resources, and to speak with authority of their finite nature. Buckminster Fuller's followers were pleased to pronounce a rigid honesty resulting in an aesthetic of logic. Foster, Rogers, Hopkins et al subscribed to an unwritten set of values. Tony Hunt shared these values, and indeed his mastery of lightweight engineering design proved indispensable during the emergence of the High-Tech Movement.

[Above] *Innovative steel fabrication techniques used to achieve the snaking, tapering plan form of the train shed at Waterloo International Terminal, a new landmark in London.*

[Opposite] *Side view of the Patera Building; press-formed panels are used for both walls and roof.*

After the mid 1980s, opportunities for Tony Hunt to work directly with manufacturers (with their preparedness to invest in specialist tooling or prototyping) became scarce, limiting possibilities for innovation. Hunt's quest for advance took a different direction. His design initiatives focused on matching more advanced and innovative fabrication techniques, allowing more emphasis on sophistication of steel working. Qualities of finishes and corrosion protection had also advanced. At Waterloo International Terminal and then at Kirklees Stadium, Huddersfield, methods such as the telescoping of joints (where a smaller diameter tube is inserted into that of a larger diameter), the tapering of tubes during the fabrication process and of course the bending of tubes brought innovations from fabrication techniques rather than from technology itself.

The last and possibly the greatest of Hunt's technological advances remains sadly only an unbuilt design. This was the highly-advanced steel structure, clad in structural glazing for the 1994 Centre du Conferences Internationales de Paris (CCIP), close to the Eiffel Tower in Paris. Architect Francis Soler demanded that there should be no diagonal cross-bracing whatsoever in the three rectilinear fully-glazed steel and glass buildings. A series of vertical and horizontal vierendeel trusses would have been used to support two skins of glass 3 metres apart both for the walls and the roof. In Paris, the Soler/Hunt design for CCIP was the preferred choice of the commissioning panel, and both parties were selected and then appointed as architects and engineers for the project. Anthony Hunt Associates set up a Paris office – headed by Nick Green – in a very short time to make themselves ready for what would have been a demanding yet exhilarating design challenge. Alas, approval for the scheme was withdrawn at the eleventh hour on governmental authority and the whole project was abandoned. However, the presentation work, as well as models and graphic representations, included another technological advance in the field of computer modelling and computer-generated image making. The designs were so far advanced that all structural elements had been designed in detail. The jointing methods and the glass carrying brackets had been designed down to the last detail. Tony Hunt had personally devised a batch-produced bracket that was the key element in linking the glass walls to the structure. This allowed super-real fully-rendered images to be produced. These have become almost commonplace over the last decade and a half, but in 1994 this technique was pioneering work. Hunt says that the models were good but the computer 'virtual reality' was better. The images produced of CCIP give life to a project which sadly belongs to a select band of 'if only' visionary architectural and engineering works. One positive outcome of events as far as AHA were concerned after the disappointment of CCIP, was the establishment of their Paris office, which thrived and prospered under the stewardship of Nick Green.

Tony Hunt recalls that on the occasion of a presentation before the clients' commissioning panel, he and architect Francis Soler, together with Nick Green of AHA, were awaiting their opportunity to present their designs. Coming out of the committee room was Peter Rice – Hunt's rival on this occasion – with an alternative

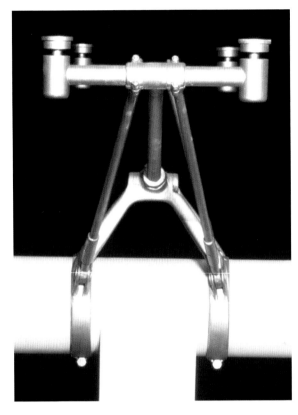

[Left] *The CCIP key component: a universal glazing bracket capable of batch production.*

[Below] *Tony Hunt's sketch of a universal bracket to meet the challenge of structural glazing for the refined steel structure of CCIP.*

[Opposite] *Aerial photograph of CCIP showing its intended site location in the shadow of the Eiffel Tower on the bank of the River Seine in Paris.*

[Previous page] *An interior CGI of the highly advanced CCIP steel structure, clad in structural glazing without any cross-bracing.*

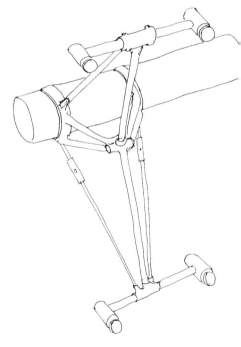

INTERNATIONAL CONFERENCE CENTRE - PARIS

FULLY ADJUSTABLE MAIN GLAZING BRACKET

team and scheme and with Arups as engineers. Hunt would have considered Rice as a friend and colleague in almost any other situation, but here they were professional rivals competing for the same job, pitting their wits against each other. It was Anthony Hunt Associates that came out on top of this particular encounter. Peter Rice together with Ted Happold had worked together for Ove Arup and Partners on the George Pompidou Centre at Beaubourg in Paris for the Piano and Rogers-designed – but some would say Archigram-inspired – building. Peter Rice – born in Ireland in 1935 and educated at Queen's University Belfast and Imperial College London – had joined Ove Arup and Partners in 1956. He became a Director of Ove Arup in 1978. Although he had worked on the difficult glazing design for the Sydney Opera House, it was on the design of the Pompidou Centre in Paris that he excelled. Between 1978 and 1980 he was in partnership with Renzo Piano – the architect who (with Richard Rogers) was responsible for the design of the Pompidou Centre. Piano and Rice took an adventurous approach to building with projects such as the IBM travelling pavilion.

The IBM Travelling Exhibition was housed in a prefabricated pre-finished pavilion designed by architect Renzo Piano. It visited 20 European cities, including London and York, in 14 countries between 1984 and 1986. The pavilion was a transparent vaulted composite structure, 48 metres long by 12 metres wide. Acting both as the glazing

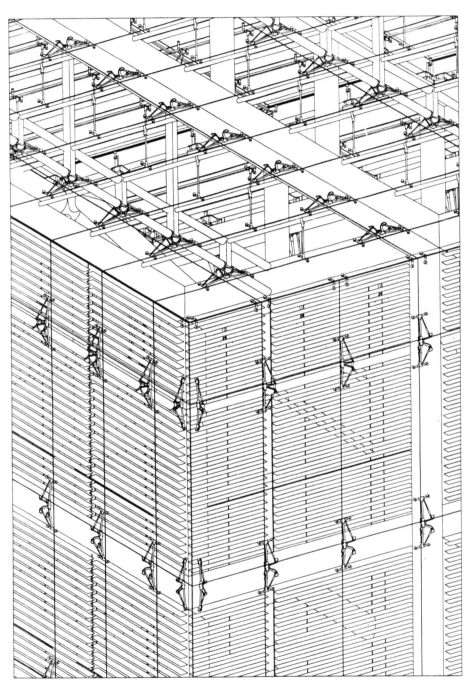

Batch-produced components would have been the abiding image of CCIP.

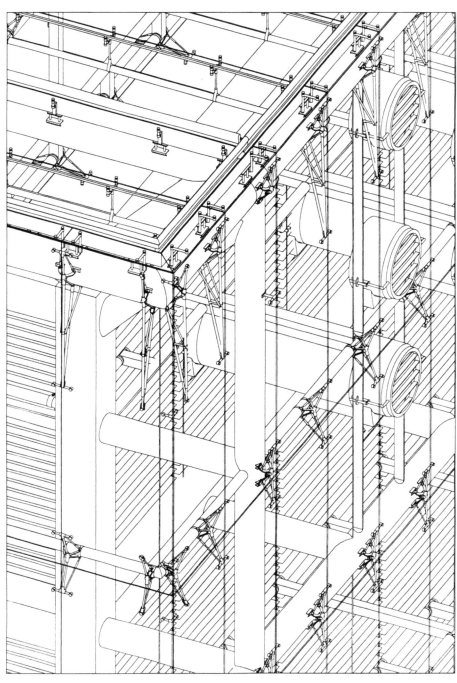

Similar brackets were to have been used both inside and outside of CCIP.

element and as part of the structural web, polycarbonate moulded pyramids were fixed together to form three-pin arched bays, 12 pyramids per bay, 34 bays making up the length of the building. The height at the apex of the vault was 6 metres. The pyramids measured 1.4 metres by 1.5 metres with a depth of about 800mm. Curved timber structural ribs ran up both the inside and outside of the polycarbonate web. No cladding was needed. The fixings were metal brackets and plates, which ensured accurate and secure joints. Dissimilar materials were bonded together using adhesive – the joints being specially prepared for this. In the design of the metal joints thermal expansion was allowed for. The pavilion had its own suspended timber floor, under which its service ducts were located. The floor was levelled using a series of adjustable jacks. The whole process of relocation took about three weeks and the components were driven from site to site in specially-built container lorries. The locations chosen were high profile – in London it was erected in the grounds of the Natural History Museum. Wherever possible wooded leafy areas were selected, which provided some natural shade for all the glazing but still made use of summer's high daylight levels to light the interior. The pavilion was manufactured to very high quality standards and finished in a way that suggested a high level of craftsmanship. However, the batch-production of identical components confirms it as a system building.

Peter Rice formed another partnership in Paris with two of Tony Hunt's former architect friends and associates – Martin Francis and Ian Ritchie. Peter Rice died suddenly in 1992. The sense of architectural adventure which had been evident in the 1960s and early 1970s in London had moved when the main protagonists such as Rice, Francis, Ritchie and Piano left the London scene.

In 1982 Tony Hunt sold Coln Manor, moved back to London, married Diana, and then very soon, he moved back to the Cotswolds to live whilst working out of the AHA Cirencester office. This was in the mid 1980s when the High-Tech Movement was floundering and Tony was no longer the automatic choice as engineer. Personal and professional relations that Hunt had enjoyed with Richard Rogers and Norman Foster in previous times had ceased to be close. After the Foster prototype house project was abandoned in 1978, Tony Hunt had received little in the way of commissions from Foster and Partners. It was a similar story with Richard Rogers. The Inmos Microprocessor project of 1982 was followed by a barren period as far as that practice was concerned. During the period from 1978 to 1984 it had been Michael Hopkins with whom Hunt had associated himself most closely. Hopkins shared Hunt's vision of component-based structures, as well as common interests in furniture, product, yacht and car design (Michael Hopkins favoured a Porsche as a suitable vehicle to be photographed against his buildings and Tony Hunt drove fast cars including a Mazda RX7 at about that time).

Tony Hunt was able to bring a fresh understanding of technology to structural design, and whereas the technology wasn't necessarily new, its transfer from other industries to building often was new. Thus his application bears comparison with Paxton or Barnes Wallis. In other industries - like car manufacturing - Modernism was starting to become influential. There was a groundswell of opinion suggesting that good design should be available to all.

In 1933 in the spirit of the Bauhaus Movement, German engineer Ferdinand Porsche (1875–1951) had a vision of a mass-produced vehicle. In 1934 Hitler backed the development of Porsche's ideas by supporting the design and manufacture of a 'people's car'. Hitler was a car enthusiast, though he was unable to drive himself. It was Hitler who influenced the design brief saying that the car should carry two adults and three children at a speed up to 60 mph, with a return of 33 mpg. The price of such a car should be 1000 Reichmarks. Ferdinand Porsche owed some of the design styling to the Czech car, Tatra. It was however the VW Beetle chassis that was used in the first of the 911 Porsche designs of 1963. In 1938 Hitler laid the foundation stone for a new factory to be known as KdF Wagen (Kraft durch Freude – strength through joy). KdF-Stadt was a new town to support the factory and was to include workers' housing, however the project was put on hold in the advent of WWII. Vehicles were made but they were for military use. After WWII, the KdF factory was in the British Quarter. KdF-Stadt was renamed Wolfsburg and the manufacturing plant was known as Wolfsburg Motor Works. Major Ivan Hirst – a British Army Officer – was brought in to assess the possibility of starting up the factory again. Contrary to advice given to him by representatives of the British Motor Manufacturers (including Lord Rootes), the factory was re-opened. The name was changed to Volkswagen (meaning 'people's car') and in 1972 the VW Beetle overtook the Model T Ford as the most popular car ever made. 20,000,000 units were made by 1981. In 1974 production at Wolfsburg was switched to the VW Golf, but production continued until 2003 (mainly in Mexico) when VW Beetle number 21,529,464 was made and taken back to the museum in Wolfsburg.

Working for Felix Samuely gave Hunt direct lineage back to Modern Movement ideals, both aesthetically and politically. Samuely - who was able to impart such passion for design to Hunt - had personal and professional contact early in his career in the UK with Lubetkin, Ove Arup and others who were to shape British Modernism. The advances in concrete technology - both practically and structurally - were to determine the ethos of the Modern Movement, just as timber woodworking had determined Paxton's work, and aluminium tube-making had determined that of Barnes Wallis.

There was a period in the mid 1980s when the High-Tech architects all turned their backs on the Movement, and this left Hunt slightly out in the cold. It had been thought that the reason was that Rogers, Foster, Hopkins etc preferred Peter Rice and therefore Arups (where Rice was a Partner) as their engineer of choice. After the design of the Pompidou Centre in Beaubourg, Rice had set up offices in Paris close to Ian Ritchie and Renzo Piano and Martin Francis had followed him there. It was Arup who engineered the Lords Mound Stand for Hopkins – this was the first big job that Hunt didn't get from Hopkins. Hunt's personal and professional relations, and what had been a close friendship with both Michael and Patti Hopkins came to an end on the occasion of the unveiling of the Lords Cricket Ground's Mound Stand. Tony Hunt has said many times that the reason for the frostiness had nothing to do with Peter Rice, with whom Hunt enjoyed a very good professional relationship. The reasons that Hunt cites are much more to do with ill-judged and inappropriate remarks made by the Hunts (both Tony and Diana) on that occasion. The falling-out was probably compounded by the fact that Diana had worked for the Hopkins' at Broadley Terrace (that was where Tony met her), and demonstrated some lack of cordiality towards her former employers. Hunt was able to mend fences to a certain extent. He continued to work with Grimshaw and Rogers. Foster and Partners eventually engaged AHA, firstly for the Sackler Galleries in 1985 (this project lasted until 1991), and for the National Botanic Garden of Wales project in 2000. Sadly however, he never worked with Hopkins again, although reasons for this might be that Hopkins moved away from exposed steelwork and he had never favoured Modernist exposed concrete structures: the two materials in which Hunt excelled. Hopkins advocated tented structures – Hunt only designed one major building with a tented roof: the Don Valley Stadium grandstand.

This period was pivotal in Hunt's career, and it coincided with the period when the face of architecture and engineering changed dramatically, due in a large part to external political influences. The emergence of Thatcherism promoted the self-importance of opinion makers, politicians, sports stars, business people, celebrities, broadcasters, journalists and architects: the perceived status of the individual became paramount. The advent of corporate monument making and the desire for landmark or 'flagship' buildings followed. What was almost a Royal Decree that city architecture should follow a prescribed pattern sealed the fate of High-Tech architecture. This was the advent of the Post-modern Movement. Gone, probably for ever, were the socialist ideals of the Bauhaus and the Modern Movement. The Sackler Galleries project illustrates this new mood in architecture. Additional gallery spaces were built on the top floor of the Grade I listed Royal Academy of Arts at Burlington House in London – originally built in 1768 and designed by architect Sir William Chambers. Hunt's technological advances in this instance were concerned with managing the structure within the confines of a Grade I listed building. Twin circular section columns 9 metres long were inserted between existing buildings allowing new landings for the two extra floors of gallery space in what had been unremarkable Diploma Galleries. The project improved circulation throughout the building by introducing lifts and staircases into what had been a light well. From AHA, Nick Green and Albert Williamson-Taylor designed a structure that was specifically tailored to suit the brief. This was a far cry from the industrialised concept of structures envisaged just a few years earlier, but what replaced the batch-produced components was a beautifully detailed structure of sculptural quality fashioned to take its place in the historic surroundings of the Royal Academy. Here again we see the ability of Tony Hunt and AHA to respond to circumstances.

[Above] *New structures introduced into the Grade I listed Royal Academy of Arts at Burlington House in London.*

[Right] *The Sackler Galleries project illustrates this new mood in architecture.*

8　POST-MODERN WATERSHED

From the mid 1980s there was a sea-change in British architecture and therefore with the contribution that Tony Hunt was able to make. By this time, architects with whom he had been closely associated had turned their backs on ideas of industrialisation, prefabrication and component-based bolt-together system buildings – ideas that had been central to their design philosophy only a decade earlier. The invaluable influences of Jean Prouvé, Konrad Wachsmann, Richard Buckminster Fuller, Charles and Ray Eames and even Walter Segal had ended with their deaths – Charles Eames died in 1978, Konrad Wachsmann died in 1980, Buckminster Fuller died in 1983 (87 years old), Jean Prouvé died in 1984 and Walter Segal in 1985. For comparison in the chronology of the High-Tech movement, Lloyds of London – possibly the most famous example of the High-Tech style in the UK – was nearing completion in 1984, and there was little to follow. Lloyd's can be seen as the high water mark of High-Tech. It is both the most advanced example in the UK and paradoxically, it is one of the last. Effectively, the dwindling influences of these key protagonists together with a changing political climate in the UK signalled the end of an era.

There are countless examples of designs and even design movements that have fallen victim to political pressures. In the 20th Century we can point to examples (already referred to above) such as the Barnes Wallis-designed R100 airship. This was a proven design, but because of shortcomings in the design of the rival government-backed R101 and the ensuing disaster of 1929, the R100 was unceremoniously scrapped. The same fate awaited the majority of buildings and structures of the 1951 Festival of Britain, when the incoming Conservative government instructed that they be torn down in 1952 just a few months after their building. The above-mentioned CCIP project in 1994 fell victim to political pressures, showing that internationally there can be similar unfortunate outcomes. Of less dramatic outcome was the Lubetkin and Tecton design for the Highpoint II flats in Highgate, North London. Although built only three years after Highpoint I (that with its impeccable Modern Movement socialist political credentials), Highpoint II was substantially altered to provide luxury high risc living due to political pressures brought about at a local level under the recently introduced Town and Country Planning Act of 1932. From a design point of view Highpoint II is an equal to Highpoint I, but from a social or political view, the two designs are worlds apart.

Political pressures from a different source were to come to the fore in the mid 1980s. The 1984 date stands out. On this occasion, pressures were brought to bear before a certain building project had even started. May 30 1984 was the date of a speech given by HRH Charles Prince of Wales to the Royal Institute of British Architects (RIBA) at a Royal Gala Evening at Hampton Court Palace, to celebrate the 150th anniversary of RIBA. The Prince of Wales said: 'What, then, are we doing to our Capital City now? What have we done to it since the bombing during the war? What are we shortly to do to one of its most famous areas – Trafalgar Square? Instead of designing an extension to the elegant facade of the National Gallery which complements it and continues the concept of columns and domes, it looks as if we may be presented with a kind of municipal fire station, complete with a sort of tower that contains a siren. I would understand this type of High-Tech approach if you demolished the whole of Trafalgar Square and started again with a single architect responsible for the entire layout, but what is proposed is like a monstrous carbuncle on the face of a much loved and elegant friend'. The Prince of Wales was of course referring to architects Ahrends Burton and Koralek (ABK) and their design for extensions to the National Gallery – designs that had been selected by competition under the auspices of RIBA itself. In his speech aimed at the heart of the architectural elite, the Prince of Wales was confirming what had always been the problem with High-Tech architecture: that is clients, patrons, funding organisations, planners and building regulation authorities could not see lightweight buildings taking their place within the fabric of the city. In the introduction to his book *Ernö Goldfinger: The Life of an Architect*, Nigel Warburton speaks of the 'Prince Charles fallacy'. This is the misguided notion that potential new buildings should only be allowed to be realised if they look more-or-less like the buildings already in the area. From this date in the UK, the search was on for a different type of architecture – one that was safe, reassuring, sustainable and uncontroversial. The era of High-Tech architecture was over. There was no place in cities for a lightweight engineering technology that owed more to aircraft, motor vehicles or yachts than it did to buildings. Indeed, any origins of High-Tech in the Pop Art Movement – as exemplified in the Archigram presentations with the attributes defined for them by Hamilton's letter to the Independent Group (with references to low cost, expendability, and gimmickry) – contrived against the acceptance of High-Tech architecture by clients and patrons.

The UK government had been under the premiership of Margaret Thatcher's Conservatives from 1979 and during the first decade of her leadership most of the Modern Movement architects' political ambitions had been abandoned. Ideas of a Utopian socialist future such as that described by Lubetkin – 'nothing was too good for ordinary people' – were replaced by the Michael Hopkins creed of 'real buildings with real sites for real clients'. There was a sub-text in this creed requiring that clients have deep pockets, thereby distancing the new mood from the Bauhaus belief: 'To serve its purpose perfectly ... to be durable, inexpensive and beautiful.' Of course Margaret Thatcher is remembered for her famous interview with *Woman's Own* magazine on October 31 1987, in which she said: 'I think we have been through a period when too many people have been given to understand that if they have a problem, it's the Government's job to cope with it. 'I have a problem; I'll get a grant. I'm homeless; the Government must house me.' They are casting their problems at society. As you know, there is no such thing as society. There are individual men and women and there are families. And no government can do anything except through people, and people must look after themselves first. It is our duty to look after ourselves and then also, to look after our neighbour. People have got entitlements too much in mind, without the obligations. There's no such thing as entitlement unless someone has first met an obligation.'

Lord Rogers had taken his place within the Establishment, demonstrating a particular interest in the fabric and development of London. He had worked on designs for the South Bank and the Coin Street scheme. He had pioneered high-rise High-Tech at the Lloyds of London building, he designed the Lloyds Register of Shipping building and his practice Rogers Stirk Harbour is currently working on another monumental high-rise tower for Threadneedle Street having recently completed buildings at No 1 Hyde Park. He was co-author (with Eton-educated Labour M.P. Mark Fisher) of a book about the urban renewal of London – *A New London* published in 1992.

> HRH the Prince of Wales intervened in the Chelsea Barracks development in 2009. Here, pressure was brought against the developers of the project, who abandoned their designs for the site. These designs were in fact the work of Richard Rogers and Partners. Richard Rogers was one of the so-called High-Tech architects who had successfully reinvented himself after the mid 1980s rejection of all that High-Tech stood for, and who had succeeded in offering the safe reassured designs that were acceptable to City patrons.

Completed in 1986, Lloyds of London houses a worldwide marketplace for the insurance industry. Its organisation dates back to the 17th Century. The simplicity of its multi-storey rectilinear plan form with a 12-storey centrally-positioned glazed atrium is obscured by a series of service towers. These create a disorderly ad hoc appearance to the building when approached from the surrounding medieval streets. The shiny staircase towers and plant room pods placed on the outside – whilst primarily there to provide servicing to the offices – become the identifying features of the building. Reminiscent of the Crystal Palace exhibition building of 1851 with its high overarching glass roof, the atrium at the heart of the building houses a series of criss-crossing escalators providing efficient access to the upper floors. It has been described as a cathedral-like space, the formality of which is advanced by the fact it houses the Lutine Bell. Salvaged from the wreck of *HMS Lutine* after it sunk in 1799, the bell was traditionally rung to herald important announcements that might impact upon the insurers, one peal sounding for bad news, two if the news was good. The architectural imagery of Lloyds of London is an 'inside out' form of design, which has its roots in the generation of architects emerging from the Architectural Association in London of the 1960s. Archigram (mentioned above and defined as a pop-culture movement involving students and tutors at the Architectural Association) envisaged 'instant cities' comprised of 'plug-in' buildings where prefabricated modules could be positioned into a structural framework. The service modules, rooftop cranes and open cellular grid-like structure of the Lloyds building replicates the imagery of Archigram.

How incongruous it is that such a pioneering design should be considered for Lloyds – the epitome of establishment values in the heart of the City of London. Initially the Lloyds members and their work people alike received the designs in horror and their hostility continued even after the building was opened. From the engineers' and architects' point of view, brave decisions were required to take what were then avant-garde and vaguely-defined concepts coupled with their experience of industrialised building methods into the world of city centre, high-rise, and high-density occupation. Many of the images, such as the plug-in modules, the rooftop cranes, the open lattice steelwork, the layout and construction of office modules (which suggest that more could be added if they were needed) have antecedents in the Archigram graphics. Yet the contrasting conservative engineering requirements of long building life, city centre complications, an exacting brief and suspicion from the client's sides (values which both made use of the building and which funded it), all conspired to act as a counterbalance to this imaginative design.

The building, therefore, is in the style of a loose fit, flexible, adjustable industrialised kit of parts, but in reality is designed and built to exacting engineering standards with little likelihood that the 'plug-in' aspects of the design would ever be utilised. However, Lloyds of London stands as one of the few examples of what is now seen as an exciting design movement centred on London in the 1960s. The watershed was reached when the drive for technological advance gave way to a new search for uncontroversial conformity – within the new boundaries of safety, sustainability and historical context, postmodern values were enforced to provide reassurance to investors.

Colin Davies in his *High Tech Architecture* book published in 1988 gives an even more precise date for the end of the High-Tech movement: January 28 1986. This was the day that the *Challenger* space shuttle blew up in front of the world's media. Failure of a component – a gasket seal not dissimilar to those seen as capable of providing a perfect seal on High-Tech buildings – was seen as the cause of the *Challenger* disaster. However, unlike the experience of High-Tech architecture, the Space Shuttle programme was re-established, allowing technological advances to rectify and overcome the failures of 1986. The same possibilities of ironing out the shortcomings of the first High-Tech buildings, which would have allowed a bank of testing and experience to be established, were never afforded to architects who had been so vocal in their praise for prefabrication and other elements of High-Tech design only a decade earlier. Lightweight structures, usually modular and sectional in appearance, inevitably became associated with low cost and low value temporary buildings. This was an association that could not be broken. Of all the High-Tech architects it was Michael Hopkins who reacted to the criticisms most strongly and moved furthest away from any association with lightweight, low value designs with his use of concrete, stonework, bronze, lead, timber and brickwork.

Tony Hunt describes himself as apolitical. He has always responded to the clients, the architects and the projects and their funding arrangements prevailing at the time of any project. His admiration of the British Modern Movement designs is restricted to the buildings themselves rather than any interpretation of their underlying political aspirations.

One architect – Walter Segal (1907–1985) – kept his belief in the Socialist principles of the Modern Movement right up to the date of his death. Born in Germany of Jewish descent but brought up in Ascone, Switzerland, he moved to London in 1936. In 1962 Segal wanted to rebuild a house that he had bought in Highgate. For the purpose of designing somewhere that he and his family could live during the construction period, he built the Garden House. In doing so, Segal pioneered a low cost, component-based timber construction method. The key elements to this were that it should be of simplified design, use standard size materials (for instance by using uncut plywood sheets), use dry joint construction (such as nuts and bolts to effect structural joints), have minimum foundations to suit its lightweight, and crucially, should combine the roles of architect, quantity surveyor and structural engineer. The self-build method that Segal developed was first tried out on his own ski lodge in Switzerland, where all components were carried to the site of the lodge by horse and sledge. He was a keen skier, making this a family-derived project. This was an unlikely first step in what was to become a community-led self-build method, sometimes called the Segal Method. It has appeared as a trait common to many of the British Modern Movement architects that they start their attempts to balance out social injustices from a position of reasonable affluence.

In 1978 a group of 14 families from the local council's waiting list, pooled their resources and under Segal's guidance built their own two-storey houses in Lewisham, south east London. A contribution to the building processes offered by Segal's methods was to demystify the technical processes of building by the use of freehand drawings, which could be easily understood by all concerned. Inevitably, the buildings produced were low cost constructions and sectional in appearance. The mid 1980s – coinciding with Segal's death – saw the advent of post-modernist architecture. The Segal Method gave self-builders a sense of worth. The Lubetkin belief from the 1930s that 'nothing is too good for ordinary people' would have provided assurance to Segal's followers. In the event, the low cost solutions and the need for inexpensive materials conspired against the wider use of the Segal Method.

Segal's contribution is worthy of mention because it fulfils many of the requirements that might be associated with the early High-Tech Movement. It is component-based, it has industrialised (though in this case not pre-fabricated) supply, it has dry construction, it is lightweight, it is quick to erect and can even be reconfigured or relocated, it is flexible and adaptable. The difference is that the components are not especially designed for their purpose and therefore are not identifiable as belonging to that particular design. Also, the low cost nature and sectional appearance of the construction has negative connotations such as 'temporary' or 'insubstantial'. As we have seen, these associations are difficult to counter as far as answering questions of funding or client approval are concerned.

Tony Hunt has referred to occasions when he was required by architects to produce new sets of details for subsequent projects, after rational and logical solutions had already been developed for given structural solutions. He refers to a 'possessiveness' of architects from the mid 1980s onwards. He says: 'Some have been prepared to use similar details proposed by me/us on a project for another architect but some have insisted – no matter how illogical and perhaps how much more expensive – that for instance, we create a new set of details for a job. The connections for Schlumberger are an example. We wanted to use castings as per Inmos but Michael Hopkins insisted on all joints fabricated out of the solid steel. One other case – Laurie Abbott, a partner of Rogers – objected to my structure for Steve Martin's house, as he said it was 'derivative' and we had a row. It never got built anyway.' Laurie Abbott had worked for Rogers since joining Team 4 when he worked on Creak Vean with them. Steve Martin was a regular visitor to the River Café, which had direct links to the Richard Rogers Partnership (where Abbott was a partner) through its location and through its founders Rose Gray and Ruth Rogers. Possessiveness would suggest a quest for the individual recognition of architects in the design of their buildings. Such a quest would take priority over the search for a tried and tested design formula, which could then be applied to subsequent projects. Possessiveness also worked against the intellectual sharing of design issues out of common interest between architects and engineers, whereas in the past, ideas would have been 'bounced around' and 'cross-fertilised' between like-minded colleagues regardless of the ownership of a particular project or design. In the earlier example of the Highpoint I apartments - designed by Lubetkin and the Tecton Group - we have seen how it was considered quite natural for other Modern Movement architects - Ernö Goldfinger and Erich Mendelsohn - to live in a building designed by another architect. Indeed, Goldfinger designed interiors for his own apartment there, including a timber sideboard supported by hot-rolled steel I-beam bearers.

One of Buckminster Fuller's last actions before his death in July 1983 was an 'open letter to architects of the world' in which, with a rallying call, he celebrated amongst other things the first solar-powered aircraft flight over the English Channel. He was optimistic because the event brought together the combination of lightweight engineering (avoiding any 'wasteful' use of materials) and renewable power, itself a move towards his dream of conserving the scarce natural resources at our disposal. By now, technology no longer offered an exciting unbridled future for personal freedoms (such as access, travel or adventure), but merely the re-enactment of a set of established images. Advances in technology became the means to make a safer, more predictable future – one in which more control was exercised over the comings and goings of mankind. Gone were the ideas that technological advances would lead to greater personal freedom or expression. Even the long-awaited manifestation of 'rocket man' at the 1984 Olympic Games opening ceremony in Los Angeles represented the end of an era – not the new dawn of some science-fiction technological revolution. By this time, UK architects with few exceptions, had abandoned their position at the forefront of technological advance. Contact with manufacturers became strictly an 'arm's length' transaction. Initiative for technological advance was handed mainly to overseas specialist manufacturers. Anything resembling system building design became taboo. The new ground rules were summed up by Michael Hopkins when he called for 'real projects with real sites and real clients'. As far as he was concerned there could no longer be any question of 'off-the-peg' building design.

In the early 1980s Michael Hopkins had confided in Tony Hunt that he faced a dilemma. Hopkins admitted that he felt more at home, more excited with, and more of a design challenge from the structures, components, materials, finishes and aesthetics on view at the Southampton Boat Show than he did with anything that the building industry could offer. He would have liked to be a part of, and to adapt the technological advances in the boat building industry for use in buildings. Hopkins spoke at length about the technology of boat building methods (including the use of extruded aluminium sections for masts by slitting, shaping and re-welding them), about glass reinforced plastic (GRP) technology in reference to advances made in the Lotus car factory in Norwich and about the stainless steel rigging components that were transferable for use in building structures. And yet, he knew that the lightweight highly-engineered, highly-finished components – such as those on show – would not be accepted into the conservative community of clients, patrons, building regulation authorities and funding organisations in the UK. Tony Hunt says that at the time, he shared this view. He too, used some of the technological advances he had seen in the boat building industry. For instance, the technique of slitting a section, shaping and re-welding as used in the fabrication of masts was used by Hunt to form tapered tubular steel elements for the curved arched trusses at the Kirklees Stadium.

Michael Hopkins expressed these views in 1980-1981, close to the end of the High-Tech era. He was deeply unhappy to see his hard-won principles of technological supremacy compromised with the advent of postmodernism. He saw straws in the wind well in advance of the period 1984–1986 when the inevitable change was there for all to see. He intimated at one stage that he would willingly give up architecture in favour of yacht design in the Mediterranean (perhaps aware that Martin Francis had made such a move). He even suggested (presumably as an affront to postmodernism) that his house in Hampstead could be refashioned in a postmodern style, possibly mock-Georgian. These were merely defiant gestures against the incoming tide of postmodernism, heralding the end of the Buckminster Fuller creed and ultimately of the High-Tech Movement.

To see how stagnant building methods have become we only need to look at the continued advances in sailing yacht technology that have taken place between the 1980s and the present. Sails that are woven on moulds to their most efficient shape instead of being made from flat sheets, and masts that can be pre-stressed into a curved shape to match the shape of the sails are just two innovations. If we look at motor vehicle technological advances, numerous developments have taken place – such as traction control, anti-skid braking systems (ABS), minimisation of wind resistance and increased fuel efficiency. Both of these industries have delivered aesthetic changes to represent those advances. Buildings from these three decades do not exhibit any similar evidence of technological advance in their structure. Building designs are certainly more flamboyant and they can be individual in their appearance, but the basics of construction remain stubbornly unchanged. In comparison to car making, the status of building construction has been likened by Michael Hopkins to the inter-war years, when cars such as MGs or Morgans were built from a chassis onto which the engine, transmission, bodywork etc were bolted. Whereas cars are now made from a monocoque lightweight shell with many technological advances, buildings remain tied to their structure in the way that those inter-war MGs or Morgans were tied to their chassis.

This dilemma of Hopkins's encapsulates the reasons for the demise of the High-Tech dreams. As we know, Hopkins and his contemporaries elected to follow the safe, reassuring and uncontroversial path dictated by their clients, patrons and funding organisations. The continuing use of the Coronation Coach can be described as a paradox, and an illustration of the dilemma faced by architects. Its construction belongs to an earlier century and yet despite the multitude of technological advances in personal transport that have been made in the intervening years, and as a part of the symbolism of the occasion, historical precedent would dictate its suitability in favour of any alternative. The form and appearance of such a vehicle has evolved throughout history. It is recognised as a symbol of State occasion. Any modernisation, such as improvements for reasons of safety or comfort, would detract from its identity. Historical precedent and enactment of State occasion provide reassurance as to the form of the vehicle. With his 'carbuncle' speech, HRH the Prince of Wales was reinforcing the views of patrons that historical context outweighed technological advances if such advances were to impact upon the traditions of city architecture, especially those of London.

Built in 1762 for King George III by Samuel Butler at a cost of £1,670, the Gold State Coach has been used at every Coronation since that of King George IV. The coach weighs almost four tons, is 24ft long and 12ft high. It is gilded throughout and features painted panels by Giovanni Cipriani. Its rich baroque style is completed by gilded sculpture including three cherubs on the roof and four tritons, one at each corner. The coach is pulled by eight Windsor Grey horses. Of course, the Prince of Wales' opinions would be vindicated by simply comparing the Samuel Butler coach with its latter day equivalent: the vehicle built to carry Pope John Paul II on the occasion of his UK tour in 1982. For this purpose, a Leyland Costructor six-wheeled commercial vehicle was adapted and amour-plated. Nicknamed the 'Popemobile', it has the registration number SCW 532 X.

The late baroque splendour of the Gold State Coach (Coronation Coach) confirming that historical context outweighs technological advance, if such an advance were to impact upon the traditions of city architecture, especially of that in London. (Image courtesy of Jim Richmond)

History has shown us how Michael Hopkins resolved his dilemma. The 1994–1995 Buckingham Palace Ticket Office provides testament and is a symbol of reassurance. This small building, rich in boat or yacht building imagery, is designed to be repositioned each spring and removed each autumn (its brief was that it should be relocatable). It is representative not of the new-found technologies of painted gel coats, GRP and aluminium – to which Hopkins referred a decade earlier in the context of yacht building – but evocative of an earlier era of stately bespoke craftsmanship, of traditional woodworking, and definitively, of wooden masts (rather than the sophisticated aluminium fabrication techniques that he had admired) – a demonstration if it were needed, of the Coronation Coach paradox, where reassurance and tradition take precedence over technological advance. The Coronation Coach paradox can be summed up as follows: architects given a brief to design will automatically look to recreate an image using historical or pre-existing forms in order to satisfy their clients and funding organisations, rather than examining technological advances to see how they might be expressed in building form. Every corporate client organisation would like a Coronation Coach of their own. They look to portray a symbol of wealth and affluence. This emotion of familiarity has to do with continuity and confidence, of historical values that are paramount in the business world. Most buildings now outlast the businesses that occupy them, but the qualities of reassurance and confidence are transferable as assets to new owners or users. The Coronation Coach is synonymous with a guarantee that nothing might be too expensive or elaborate for its position. This sentiment is in contrast to Lubetkin's opinion that 'nothing is too good for ordinary people', to Gropius' belief, 'to serve its purpose perfectly ... to be durable, inexpensive and beautiful', or Hamilton's list of attributes associated with Pop-Art (particularly low cost). From the mid 1980s onwards, architects felt that they must create a symbol of wealth for their clients, not a symbol of inexpensive rationale. Later claims that timber was 'from sustainable sources' and claims of other low energy or low carbon design solutions were attempts to justify a distortion of Buckminster Fuller's creed of lightweight engineering that would avoid waste and make the best possible use of the Earth's scarce resources.

The Inmos Microprocessor factory in Newport, South Wales completed in 1982 and designed by Richard Rogers Partnership represented the last vestige of the High-Tech Movement. This building was actually designed after the Lloyds of London project by the same architects, but it was completed before Lloyds. Exposed lattice steelwork pin-jointed together with expressed tension rod support and externally positioned service pods were all an echo to the mantra of Archigram and the Architectural Association in the 1960s and 1970s for loose-fit, plug-in, industrialised, adaptable buildings. Tony Hunt was left to resolve engineering designs for the Inmos project, when the spirit of architectural adventure in which it had been conceived had been all but exhausted. Hunt's friendship with Iann Barron (the Inmos CEO) helped with the commission, and with Tony Hunt's steadfast engineering input, the project was a success. Hunt had first become friendly with Iann Barron in the mid-1970s at the time of the Willis Faber Dumas project. Barron was head of a computer firm Computer Technology for which Hunt worked on a project. Artist Ben Johnson – a graduate of the Royal college of Art – was commissioned to work on a super-realist painting of the central spine of the Inmos building. He also produced a painting of the ground floor of the Willis Faber Dumas building in Ipswich. Both of these paintings celebrate Tony Hunt's structures. Possibly the best example of High-Tech architecture as far as its marriage with a high-tech client operation is concerned, the Inmos Microprocessor Plant was built for the emerging microprocessor 'chip' manufacturing industry (known then in the 1980s as a new-dawn industry). With the new building type came the architectural requirements of clean air and flexibility in layout, which required wide uninterrupted spaces that could be sub-divided and provided with very high levels of servicing. Inmos Limited was founded in 1978 by Iann Barron (a British computer consultant) with Richard Petritz and Paul Schroeder (both American semiconductor industry experts). Funds of £50 million were made available from the UK National Enterprise Board. In 1984 Thorn EMI made a bid of £124 million for the state's 76% interest in the company. In total, Inmos Ltd had received £211 million from UK taxpayers, but never became profitable. In 1994 the Inmos brand name was discontinued. The exacting brief called for rapid building methods and possibilities to extend and greatly increase its size. Hunt's response was to design a bolt together structure in enormous sections (that is as large as could be transported and lifted into position), which were connected together using stainless steel pins for the joint assembles.

In the tradition of Archigram and of Michael Webb's Architectural Association design studies, Richard Rogers's Inmos designs represent the ideals of dry construction, plug-in, externally serviced, systemised accommodation modules. The central spine – the main access corridor for the building – is like an internal street. Above it, supported by a series of lattice steel towers, are double-storey roof level service pods and plant rooms. Highly visible exposed ductwork and pipework feed into the ceilings of the production areas, supported from above by the lattice steel prismatic structure. Unlike earlier and more experimental High-Tech buildings such as Foster Associates' Sainsbury Centre for Visual Arts in Norwich, Inmos relies upon traditional flat roof technology and upon proprietary glazing systems to construct modular walls of glass, solid infill or louvred panels. The design did not provide the opportunity for technological advances in structural panel construction. The central spine is 7.2 metres wide and 106 metres long. A series of similarly sized (six bays deep) accommodation bays are positioned each side of the spine at 13 metre centres, 36 metres deep. To the north side are the production areas and to the south side are the canteen, office and ancillary areas. The design allowed for unlimited extension to the same pattern. Anthony Hunt Associates devised some complex connections for the structure, including pin joints, forked connectors and specially-designed bracketry. The principle was to bring the largest possible fabrications to site and to erect them quickly and easily. Three-legged CHS tower columns each side of the central spine provide the primary support structure. Stainless steel pins were used instead of bolted connections. The prismatic roof trusses are supported by tension rods, paired to match the twin outer members of the three-legged tower columns. Twin tension rods support the prismatic trusses at the one third point of their span, with a single rod at the two thirds point. External splayed tension

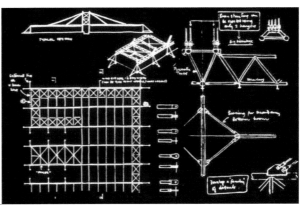

[Above] *Tony Hunt's sketches of the Inmos structural concept.*

[Left] *Inmos Microprocessor Factory, the spinal corridor an internal street.*

[Below] *Roof-level service pods, ductwork and exposed structure of the Inmos Microprocessor Factory.*

[Opposite] *A super-realist painting of the central spine of the Inmos building by Ben Johnson.*

[Left] *External splayed tension struts, held to the ground by anchor piles.*

[Below] *The impressive translucent tented roof of the Schlumberger Research building supported by an elaborate arrangement of tension cables. (Image courtesy of Iqbal Aalam)*

[Opposite] *The Schlumberger Research building in Cambridge – designed by Michael Hopkins and Partners. The spectacular tented roof pointed the way for future high-budget landmark buildings.*

struts – held to the ground by anchor piles – are expressive of the need to balance the bending moment at that point. They provide lateral bracing and support the roof trusses at their end, away from the central spine. Steel castings were used where a large number of repeat components were required. The quantity justified the investment in casting patterns and foundry processes.

From the heights of the High-Tech Movement, within a short space of time a watershed was reached. There were a few remaining projects on which Hunt was working that were trapped in the High-Tech age. By now they represented an outmoded style. The Schlumberger Research building in Cambridge designed by Michael Hopkins Architects used exposed steelwork and profiled steel cladding – techniques belonging to an earlier decade. However, the spectacular tented roof pointed the way for future high-budget landmark buildings – a new architectural movement we might describe as Tech Modern. The building was a research facility for the oil-exploration industry, housing a simulated drilling rig plus laboratories and offices. The impressive translucent tented roof was the largest of its type at the time of construction. The centre was built in two phases. The test station came first and was completed in 1985. For the unimpeded operation of the equipment planned for it, a minimum height of 10 metres and a clear span (no columns) were required. This part of the building consists of three 24 metre by 18 metre bays under the landmark tented roof. The working test area uses two of these bays. The third houses a winter garden and refectory. Flanking the three tented bays are a series of cabins – five each side. Each of the cabins is 18 metres long and 12 metres wide. External tubular steel columns and steel trusses support their flat roofs and glazed walls. The placement of the steel structure on the outside enabled the engineers to satisfy the fire-resistance regulations, since they could count the building fabric as protection for the steelwork should fire break out inside the building. This technique had been introduced by Ove Arup & Partners in the engineering design of the Patera Building by architect Michael Hopkins, 1980. (Ove Arup and Partners were appointed purely to consider the fire performance of the structural steel cladding for the Patera Building. Arup by that time had a specialist section concerned with fire safety, headed by Margaret Law). The spectacular tented roof, gleaming white by day and glowing at night, stands independent of the main structure. It is made of Teflon-coated fabric, which has a life expectancy of 30 years and is supported by its own tension structure of cables and rods. Prismatic portal frame trusses form a valley and continuous mounting anchorage for the tented roof. They also support the twin struts that direct the tension rods and enable a connection with the ground outside the line of the office cabins.

In 1989 within this uncertain phase of Hunt's career (during which he formed a partnership with YRM, [formally architects Yorke Rosenburg & Mardall]), came the

[Left] *The Schlumberger Research building in Cambridge – the building consists of three 24 metre by 18 metre bays under the landmark tented roof. The working test area uses two of these bays. The third houses a winter garden and refectory.*

[Opposite] *A super-realist painting of the Waterloo International Terminal by Brendan Neiland*

commission for the International Terminal at Waterloo Station in London – the first home of the new Eurostar service using the Channel Tunnel. Working with architects Nicholas Grimshaw and Partners, the skills that had so endeared Tony Hunt to architects in the past were once again in demand. His industrial design abilities were needed as well as his structural engineering skills. Far from being an off-the-peg design solution, this was an individual design of sufficient size to warrant the setting of manufacturing processes just for the one building. Artist Brendan Neiland also produced a super-realist painting of the Waterloo International Terminal which was presented to Bob Reid (Head of British Rail) to celebrate the completion of the project. The work is 12ft wide by 5ft high.

In 1993 the Waterloo Station International Terminal opened, heralding a new phase of rail travel in the UK. Designed as the Central London hub of the Eurostar service – a high-speed rail service to Paris and Brussels using the newly completed Channel Tunnel – the extension at Waterloo Station was thus a symbol of today's engineering, required to become one of London's landmarks. Designs underwent considerable changes from the early proposals. There are five platform viaducts supported by a grid of cylindrical concrete columns emerging from a car park at basement level and up through public circulation areas at intermediate levels, supporting the uppermost platform levels and the train shed itself. There is a structural glass screen separating the old station complex from the new. The spectacular train shed enclosure is some 400 metres long and is supported by 36 trusses of spans varying from 32.7 metres to a maximum of 48.5 metres. The structure curves in two directions and narrows towards the outgoing end. Logically, this allows wider platforms closer to the station

[Above] *View over the roof-tops, showing Waterloo International's Paxton derived ridge and furrow roof construction threaded into the established station complex. (Image courtesy of Igor Litinsky)*

[Left] *The Technique of curving and tapering prismatic trusses using advanced steel fabrication techniques at Waterloo International: tapering of tubes, telescoping of smaller sizes into larger ones, and bending to form prismatic, arched trusses. (Image courtesy of Alisdair Anderson)*

[Opposite] *The new train shed at Waterloo International is connected to the existing station buildings by a wall of structural glazing.*

entrance and exits where the passenger density is greatest. Curvature of the roof, which is in essence a flattened arch, climbs more steeply on the western side where the trains pass closely by the glass wall. The three-pin arched structure comprises two dissimilar trusses, triangular in section and having compression booms of tubular steel (CHS) and tension booms of solid steel rods. The asymmetrical nature of the structure dictates that the heavier and longer span of the three-pin arch (the eastern part with a solid roof) has two compression booms located uppermost on the outer side of the truss, and a single tension boom below. This configuration results in horizontal thrust through the pin joint onto the shorter trusses on the western side. The bending moment patterns are reversed in the shorter trusses, where only a single compression boom is required inside with two tension booms on the outside. The diameter of the compression booms vary in accordance with the bending moment from greater diameter and wall thickness where the bending moment is at a maximum, scaling down to smaller sections closer to the pin joints. This is achieved ideally by telescoping one tube into another of larger size, but due to incompatibility of tube sizes, a similar effect was achieved by plating the end of the larger tube and butt-jointing the reduced diameter tube. Also the effect was achieved by slitting tubes and re-welding them to form tapers during the fabrication process. As the sizing of the compression booms responds to the bending moment, so does the configuration of the truss, increasing in dimensions – the trusses become wider and deeper at the centre of their span, with both compression and tension members curving back to adjoin at the pin joints. Tony Hunt describes these as 'banana shaped' trusses. This technique of curving and tapering the prismatic trusses was later used to great visual effect at the Galpharm

Waterloo International

Stadium (formerly the McAlpine or Kirklees Stadium) at Huddersfield. A secondary tubular structure of CHS steel provides line bracing between trusses and offers support to the cladding and glazing. This secondary structure is in turn cross-braced by steel tension rods with forked connectors derived from yacht rigging components – a style Tony Hunt used in earlier projects. On the western shorter span of the three-pin arch, glazing with traditional overlapped transverse joints is held by aluminium sash bars slotted to reduce their weight. These are fixed to the secondary tubular line bracing on the line of the inside of the main trusses. The long eastern span has solid stainless steel panelling between the trusses – this time on the outside – on the compression booms, with glazing following the tapering shape of the twin compression booms of the trusses to provide visual relief to the solid part of the roof. There is a transverse pitch with its apex at the midpoint between trusses on the glazed western side, designed to throw rainwater towards collection points running between the bases of the trusses. On the eastern long span, the transverse pitch of the solid roof allows rainwater to flow to points centrally placed between the trusses, providing a herringbone pattern to the cladding. This is an echo of the rhomboid shapes (a technique used to allow flat panels to be formed into a curved plane) in the Great Conservatory at Chatsworth House designed by Decimus Burton and Sir Joseph Paxton. Adjustable fixing brackets and flexible glazing gaskets allow variation, sufficient for the curving tapering plan shape at Waterloo to be accommodated within a strict rectilinear structural system. This flexibility was required for a second purpose. It provided a cushioning effect to prevent shock waves generated by incoming trains being transmitted to – and thus damaging the glass elements of – the shed roof.

The result is a train shed to surpass any precedents, but which pays homage to the engineering of the railway age. By locating all other functions such as ticketing, security, passport control, waiting areas, concourse areas, arrivals and departures

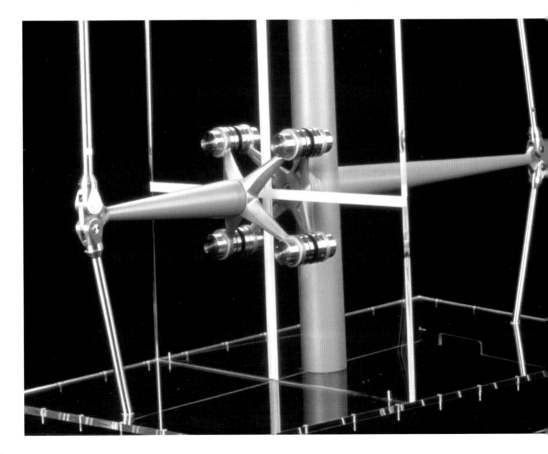

[Above] *Pin-joint at the base of the three-pin arched structure of Waterloo International.*

[Right] *Steel casting for the base joint connection prior to the fabrication process.*

[Opposite] *Detail of structural glazing between existing building and the extension.*

[Above] *Adjustable fixing brackets and flexible glazing gaskets allow variation sufficient for the curving tapering plan shape at Waterloo to be accommodated within a strict rectilinear structural system.*

[Opposite] *CGI of the vaulted span of the type envisaged in AHA's feasibility study for a new train shed at Waterloo International.*

at lower levels, the light, bright train shed becomes a track-side oasis simply for intermittent passenger use and provides a tranquil, unhurried celebration of the engineering excellence associated with the Channel Tunnel project. The structure of the train shed is testament to Tony Hunt's unique expertise in combining batch-produced industrially-designed structural components with an architectural vision. Only a handful of differing varieties of components were used, yet the result was in the tradition of Sir Joseph Paxton – unitised without being uniform, uncomplicated without being crude or uninspiring, and a logically-resolved rectilinear structural grid able to accommodate the complex snaking, tapering plan form demanded by the railway. The long snaking tunnel-like glazed roof was a triumph of design. Just a few variations of components produced in large batched quantities were assembled into the most elegant structure. Hunt demonstrated that his ideas of batch production of structural components could be used to create economic and appropriate structures for high value prestigious buildings. This was the technique that Joseph Paxton had introduced nearly a century and a half earlier for the Great Conservatory building of 1837–1840 and 1849. Technology may have changed from steam-powered woodworking to sophisticated prefabrication and highly-finished high-grade steel, but the elegance of the resultant designs can be directly compared.

Hunt was appointed as engineer for the Waterloo International project before Grimshaw had been appointed as architect. He had been asked to comment on a scheme that had been prepared by the British Rail architects department, which he says he found 'over complicated - oh and painted pink!' Therefore, Hunt carried out a feasibility study for a new train shed for British Rail architects. He also helped them compile a shortlist of architects suitable for the project which would include Grimshaw. In 2008 the high speed rail link (designed to complement the French equivalent) was completed from the Channel Tunnel to central London. However, that opportunity was used to regenerate the St Pancras station site, leaving Waterloo International Terminal obsolete after only 15 years.

The decision to amalgamate Anthony Hunt Associates with YRM proved to be a definitive moment in the career of Tony Hunt. On one occasion in early 1989, Hunt was on a skiing trip to Austria. He took a phone call from Tim Poulson – a Director of YRM – who in turn was also about to go on holiday. An offer was made there and then for AHA to be bought out by YRM and their activities to be accommodated within the small-but-expanding structures department headed at that time by Bjorn Watson. There were to be five different sections to the YRM group, of which structures was one. As with many aspects of his life, the decision to amalgamate with YRM was never going to be straightforward. It was interrupted by two similar competing offers. The first of these was from a building services engineering company. This offer was rejected but an approach from Michael Aukett Associates left Tony Hunt with a dilemma upon which to ponder. Aukett had a small engineering group, which could be expanded. He also had a thriving architectural practice – indeed AHA had worked with Aukett's staff on several successful projects. Hunt had to decide upon a future more widely based than just structures. In the end, Tony Hunt put both options to a vote before the six members of the board of AHA. By a vote of five for and one against, the decision was taken to join forces with YRM. The decision left Tony Hunt with some regrets about letting Michael Aukett down. They were friends, having worked together on several projects including work for the Sun Alliance insurance group. Hunt enjoyed visits to Aukett's offices in the Corner House, Cheyne Walk, Chelsea. Hunt refers to meetings in the top floor flat at the Corner House with Michael Aukett and an accountant friend who Aukett was later to marry. Hunt also recalls carrying out a structural survey urgently for Michael Aukett on the former home of Bridget D'Oyly Carte – Shrubs Wood at Chalfont St Giles in Buckinghamshire. This was an art deco building by architect Serge Chermayeff. The reason for the urgency was that the property was for sale by auction. Tony Hunt gave his opinion, and Michael Aukett duly purchased the house.

The importance of Tony Hunt's role at the forefront of YRM Anthony Hunt was demonstrated in the commission of the practice to work on the Eden Project. This came about as a result of a meeting at Falmouth School of Art between Hunt and a local architect, Jonathan Ball (initially one of Tim Smit's partners on the Eden Project). Tim Smit was the driving force in the realisation of this ambitious project. He had undertaken a restoration of the Lost Gardens of Heligan near St Austell, Cornwall. From this experience, Tim Smit was able to combine such high levels of technological and design input with an innate commercial ability to build the unique Eden Project.

The partnership with YRM came to an end in 1997 when the group as a whole failed. Tony Hunt was able to buy back AHA and continued to specialise in structural design, distancing himself from the rest of the YRM activities.

Hunt had made many friends in Cornwall including architect John Crowther, who with his Swedish wife, lived in a house of Scandinavian design not far away from Pill Creek at Restronguet. This was not one of Tony Hunt's own jobs, although he admired the design. John Crowther was one of a number of architects and designers that Tony Hunt met when he was lecturing at Falmouth College. On one occasion, another

An early CGI of the Eden Project showing an early, superceded ridge-and-furrow glass roof design.

mutual friend and architect from this group, Jonathan Ball, came up to Hunt after his lecture saying that 'they', meaning Tim Smit, wanted to re-appoint the design team who had been responsible for the Waterloo Station project. This was at the early stages of planning for the Eden Project. It can be said that for both Waterloo Station and for the Eden Project Tony Hunt was signalled out as structural engineer before the appointment of Grimshaw as architect

9 THE CHANGING FACE OF ENGINEERING

The role of the engineer had changed in response to new architectural requirements and as high-profile projects demanded a multidisciplinary engineering service. This would now include engineering of facade performance and energy use, fire and safety issues, all types of mechanical services and infrastructure construction as well as structural engineering. Tony Hunt was in a position to accept the new challenge.

Particularly with energy systems, but also with many other areas of building technology, architects at this stage were relying upon technological developments that were outside their own control and in many cases completely outside their sphere of influence. Their expertise and design processes now centred upon understanding advanced technology before including the appropriate equipment or systems into their building designs. It had become unrealistic for an individual architect to fully understand all aspects of building technology. A multidisciplinary team with each member contributing his or her own speciality to the design replaced the older methods that relied upon individual endeavour.

Parallels can be drawn with the advanced engineering of the aerospace industry. However, the integration of technologically-advanced systems into building shells that have changed very little over the decades indicate a lack of progress towards the ideals of the Modern or early High-Tech Movements. Acquiescence by architects to the conservatism of clients, patrons and funding organisations have restricted technological advance. Whereas nearly all of the parts may show technological advances, these advances are not necessarily evident in the design of the whole building. This has ultimately led to the hermetic sealing of buildings, with reliance upon mechanical services instead of natural ventilation.

Architectural design was now concerned with how a building looked instead of how it worked or – sometimes – even how it was built. Technology from outside sources could ensure any building would work. This change was partly due to the absence of distinctive functions for buildings – desk based activities, storage or display covered most spatial requirements. Processes have become more compact and activities have become less onerous as far as design is concerned. It was no longer easy to discern the function of a building from its design. Indeed, the function was no longer the critical factor of design. The changing face of engineering was to support the evolution of a series of new iconic monumental buildings, giving landmark buildings and enhanced status to client organisations.

The Museum of Scotland designed by architects Benson and Forsyth in Edinburgh was just such a project. Complex, intricately planned, and multi-levelled, with a towering atrium and glazed roof, the stone-clad concrete structure needed to be tailored to a steeply-sloping site in the heart of Edinburgh. Benson and Forsyth had won the competition for the design. It is said that they researched the project more thoroughly than their competitors, even to the extent that they identified which quarry they would source the stone from and that there was sufficient stone there for the project. Tony Hunt worked on several of the entries for the design competition held to appoint architects for the National Museum of Scotland. Apart from the much acclaimed winning scheme by Benson and Forsyth, Hunt recalls another by David Chipperfield that he felt would have been a worthy winner. David Chipperfield was another architect who had worked for Foster Associates before establishing his own

[Above] *The complex intricately-planned and multi-levelled National Museum of Scotland in Edinburgh. (Image courtesy of Mason Taylor)*

[Opposite] *The distinctive round tower of the National Museum of Scotland.*

practice. That Hunt was able to work on several entries for any one design competition bears testament to his versatility and his willingness to adapt his engineering skills to suit all types of architecture: the 'unseen hand' as he likes to call his design influence. Even with the most complicated design brief, he has been able to focus on developing designs for a given set of circumstances (that is differing structural systems for differing architects), without any preferences he may hold impacting upon other architects' designs. He would rarely say which design he preferred other than that he favoured component-based designs. Similarly, the Lloyds Register of Shipping building by Richard Rogers Partnership (a building that was shortlisted for the 2002 Stirling Prize), was another challenging brief. Here, steel and glass towers were introduced into the existing street frontages in the heart of the City of London. In additon, the Sackler Galleries project – part of the Royal Academy in London – required complex multi-storey steelwork to be introduced into an existing building layout.

In the 1990s a renewed building type emerged: sports stadia (mainly due to improvements for spectator safety, made necessary as a result of the Hillsborough and Bradford stadium disasters). Stadia gave more scope for the highly visible, sophisticated steel structures from which Tony Hunt had made his name. The Don Valley Stadium in Sheffield – a tented grandstand structure built for the 1991 student games – was engineered by Hunt for the Sheffield City Council's Design and Building Services Department. The design represented an advance on the cable stayed steelwork designs of Schlumberger several years earlier. For the McAlpine Stadium in Huddersfield completed in 1994, Hunt designed enormous arched lattice steel trusses to span the length of the spectator stands parallel to the pitch, affording unobstructed views to the terraced seating. Urgent reviews of safety standards in stadium design became necessary in the 1980s following a grandstand fire at Bradford, a wall collapse at Heysel in Germany, and a football match where spectators were trapped by security fences at Hillsborough, Sheffield in 1989. After looking at crowd safety, the Taylor

THE CHANGING FACE OF ENGINEERING

143

[Above] *Splayed structural supports at all levels, this photograph is taken at basement level.*

[Opposite] *The steel and glass towers of the Lloyds Register of Shipping were introduced into the existing street frontages in the heart of the City of London.*

Report called for all-seater stadia and this, together with increases in funding for sports grounds, provided an impetus that was to spawn the programme of the building of many new stadia during the next two decades.

First amongst these was the Kirklees Council's proposal for a new stadium in Huddersfield. Architects the Lobb Partnership anticipated this need with a schematic proposal: 'A Stadium for the Nineties', designs for which evolved in partnership with the Sports Council, and were unveiled at the NEC Birmingham in 1990. Kirklees Council therefore approached the Lobb Partnership and they entered a limited design competition for a 25,000 seat stadium to be built on a 50 hectare reclaimed brown-field site in Huddersfield, bounded by the river Colne on the west side and by Kilner Bank – a picturesque wooded bank – on the east. The Lobb Partnership proposed a stadium of plan form generated by limiting the distance of any spectator to 150 metres from any of the four corners of the pitch, and ensuring that every seat was within a 90 metre arc drawn from the centre of the rectangular playing area. This geometric process led to four steeply-banked segments logically giving a greater

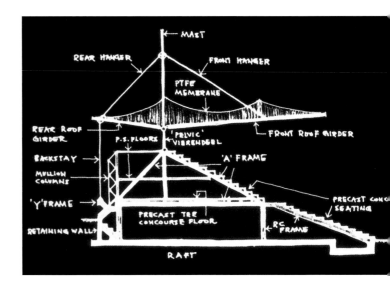

number of seats towards the centre of each stand where spectators prefer to watch from, tailing off to fewer numbers at each corner. This layout has since been criticised for lacking the spectator involvement of a true arena. Lobb recommended that terrace steps should measure a minimum of 800mm. They also created a 3-D computer modelling programme capable of affirming lines of sight and unrestricted visibility from each seat in the stadium. The McAlpine Stadium became the 1995 RIBA 'Building of the year' Prize winner.

Instead of using the accepted cantilevered roof supports radially positioned over spectator stands, the defining feature of the structural design is the use of curved prismatic trusses that support a slender roof of metal decking. Arched in line with edges of the pitch and nicknamed 'banana trusses', the technique for these long-span pin-jointed (in this case two pin) arches was first developed by Anthony Hunt Associates at the Waterloo Station International Terminal. Triangular in section and made from CHS steel tubes, the trusses have two compression boom uppermost. Struts

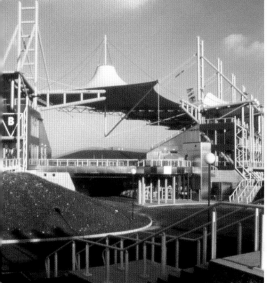

[Opposite page] *Hunt's sketch of the structural concept for the Don Valley Athletics Stadium.*

[Top of page] *Front elevation of the Don Valley Stadium.*

[Above] *View from underneath the fabric canopy of the grandstand.*

[Left] *Corner shot of the grandstands of the Don Valley Stadium in Sheffield.*

angle down to a single tension boom, from which the roof is suspended. As well as being curved into an arch, the trusses are tapered from maximum dimensions at mid span where the bending moment is greatest, to pin joints – a single one at each end. At each node point on the tension boom, the roof is punctured by connectors which fasten a series of transverse secondary castelated beams to the trusses (this is a fabrication technique where standard hot rolled universal beams are cut along their web to a pattern and re-welded to increase their structural depth and to reduce their weight). From the line of the prismatic trusses, the castellation is tapered into a cantilever to support the innermost edge of the roof. The outermost edge of the secondary beams is supported by the outside walls of the stadium. Cold rolled steel purlins allow fixing of the metal decking roof between the secondary castelated beams. The largest trusses extend to 140 metres and each weigh 78 tonnes. At each end the arched trusses are supported by distinctive concrete thrust blocks, positioned one at each corner of the stadium. These consist of square concrete platforms supported by legs angled outwards and downwards. They also provided support for floodlighting masts. In the north western corner of the ground there is a cylindrical control and security installation, made by cladding the thrust block with glass panels. Each of the four arched trusses are different, having rises in accordance with the depth and height of the stand below and by making use of the natural banking provided by the slope of Kilner Bank. Nonetheless they provide a unified appearance. The main Riverside Stand has two tiers with a row of 26 executive boxes separating them. Opposite, the Kilner Stand is a large single-tiered stand holding over 7,000 seats. The South Stand seats over 4,000 spectators. However, the North Stand is the most recent and the tallest, housing two tiers of seating separated by 16 hospitality boxes and other special viewing areas. In the lower tier, the seating is temporary so that it can be removed for other events such as concerts to take place.

[Right] *The distinctive concrete thrust blocks between stands of the McAlpine Stadium, supported by legs angled outwards and downwards.*

[Opposite] *The two-pin arched prismatic 'banana' truss of the McAlpine Stadium.*

The Kirklees Stadium pointed the way for stadium design and construction to be a component of community regeneration. It was originally jointly owned by Kirklees Council, Huddersfield Football Club and Huddersfield Giants Rugby League Club (20%, 40% and 40% respectively). Alfred McAlpine Construction Ltd – the stadium builders – entered into a ten-year sponsorship deal for the name of the stadium, but in 2003 it was renamed the Galpharm Stadium by the new sponsors. Architects the Lobb Partnership merged with the American stadium designers HOK in 1999.

Grandstand viewed from the pitch of the McAlpine Stadium.

10   LIKE MINDS

When Julian Hunt left school in 1974, with Tony having paved the way, he went to work with Michael Wickham to study furniture design. It is worth including extracts from Wickham's obituary written by Sir Terence Conran for *The Independent* newspaper, dated February 7 1995:

> 'Michael Wickham was a master of many trades and an enthusiastic amateur of countless others. He was a furniture-maker, a gardener, a photographer, a painter, a musician, a cook, a motor engineer, a raconteur, a linguist, a romantic philanderer, a Marxist, a husband (four times), a father (of seven children), a grandfather (of 12) and, best of all, an enormously generous *bon viveur*. Despite, or even perhaps because of this, he was seriously impecunious for most of his life. He was an inspirational man who had a great influence on many people's lives – a particular influence on mine; he enchanted practically everybody who came under his spell. His generosity was both material and intellectual; he wanted to share his knowledge, his ideas, his dreams and passions with everybody he met. He also wanted to feed them, ply them with home-brewed beer and home-made wine, give them plants from his garden and things he had made in his workshop…Michael Wickham was the son of middle-class parents and went to Marlborough College and then on to the Royal Academy School. His first wife Peggy Earnshaw, was the daughter of illustrator Mabel Lucie Atwell. He lived in a henhouse in Cassis in the South of France with his second wife, Tanya. It was a wonderfully creative and Bohemian environment, rubbing shoulders and palettes with Braque, Picasso and Varda, and back in London, with the artist Julian Trevelyan who became his great friend. I have always found Wickham's paintings to be practically indistinguishable from Trevelyan's. They depict decent and charming English semi-abstract views of peaceful landscapes and assemblages of fine objects: happily undemanding pictures of a better sort of world…
>
> …The interiors of his houses and workshops, culminating in the laundry of the demolished Coleshill in Berkshire, were, and indeed are, wondrous places. It would be difficult to imagine anything more chaotic. They are the work of an aesthetic magpie collecting objects

of rarity, junk, sentimental memories, engineering curiosities and the odd furry pie cooked several months before and forgotten in some corner of a cupboard…'.

The artist Julian Trevelyan (1910–1988) was a tutor at the Royal College of Art from 1955–1963 at about the same time that Hunt lectured and tutored there; Trevelyan was awarded fellowship of RCA in 1986.

Tony Hunt has referred to the house mentioned in the aforementioned extract – Coleshill House in Berkshire – as a country house with an admirable architectural heritage. Built in the 17th Century, the design owes much to architect Inigo Jones, his pupil John Webb, and finally to Sir Roger Pratt (1620–1684). Although the house itself was demolished following a fire in 1952, the laundry block where Wickham lived and worked maintained the history of the house, albeit in a small way. That Julian Hunt worked here would enthuse Tony Hunt – as, in Wickham – he saw a reflection of his own passions for life. Tony Hunt's own splendid refectory table – at which many meetings take place (in preference to an office or boardroom environment) – was made by Julian during his time with Michael Wickham. Julian adopted Wickham's furniture-making style by expressing the repairs of knot holes or other blemishes, as beautifully machined plugs and patches blended perfectly into the table top.

It was purely by accident when he was aged 22 years old, that Tony Hunt first met Michael Wickham. Tony was looking for a car to race and he answered an advertisement to look at an Austin 7 Ulster. The car's owner – Peter Brenchley – showed Hunt the car, which was in a workshop at the rear of Brenchley's property in Chiswick, west London. Above the workshop was a loft studio where Wickham had his studio. Wickham himself was a car enthusiast, owning a Bristol and a Lagonda. Wickham lived in Kentish Town at that time. Hunt drove various high performance cars including a Fiat

Tony Hunt's kitchen with the acacia wood table made by Julian.

Barchetta and his favourite, a 1975 Fiat Dino which had a 2.4 litre V6 Ferrari engine. Hunt maintained a friendship with Wickham that continued with a sort of reciprocal *in loco parentis* arrangement. Julian Hunt at the age of 15 decided that he had attended school for quite long enough and, out of frustration, Tony asked him 'what it was instead of school that he wanted to do?' In reply Julian announced: 'I should like to work for Michael Wickham'. And so, with Tony's backing, when Wickham bought the 'Laundry', Julian moved there with him. In parallel and at the same time, Gemma Wickham – Michael's daughter with his third wife Cynthia – moved from the laundry studio at Coleshill to London in order to study at secretarial college. She lived with the Hunts during this period of study. By coincidence, when Gemma finished at college, the AHA move to Coln Manor was imminent and so she moved with them to the Cotswolds to become a part of the permanent staff there.

The last of Tony Hunt's six decades in the profession provided some outstanding commissions. These opportunities partly arose due to exceptional funding made available in celebration of the millennium, but also because of Tony Hunt's ability to search out like-minded individuals, as he had in the formative years of his professional life, both architects and friends.

One such example of Hunt seeking like-minded individuals is his renewed acquaintance with James Dyson (a former pupil of Hunt's at the Royal College of Art, now Sir James Dyson). This was during Dyson's search for new premises in which to accommodate his flourishing business in high-profile vacuum cleaners, which used the Dyson-patented cyclone technology. Hunt and Dyson soon became friends. The search for premises ended with an existing factory building at Malmesbury, Wiltshire being converted in 1996 by architects Wilkinson Eyre – one of three architectural practices that Tony Hunt had introduced to James Dyson. They became lead designers with Hunt and AHA assuming the normal support role as structural engineer. It is worth noting that after spells working with Norman Foster, Richard Rogers and Michael Hopkins, Chris Wilkinson set up his own practice in 1983. Four years later he was joined by Jim Eyre to form Wilkinson Eyre. The practice specialises in sophisticated steel structures, including the Millennium Bridge in Gateshead. Wilkinson was elected Royal Academian on March 27 2006. Tony Hunt recalls that the Malmesbury factory for Dyson was being built at a time of very rapid expansion in the Dyson business. Phase 1 led into Phases 2 and 3. Soon the site was full, leading Dyson to look elsewhere for production facilities. Of note was a furniture system that James Dyson himself designed, capable of being produced quickly and economically to keep pace with the rapid expansion and the corresponding increase in personnel.

In 1994 Hunt sought another like-minded individual in the renewal of the personal co-operation between he and Jan Kaplicky, in the designs for the Hauer King House in Canonbury, London. The clients were Jeremy King (co-founder of The Ivy restaurant where Tony Hunt met with him) and Debra Hauer. The rear of the house features a long sloping roof of structural glazing. One image of this design demonstrates the innovative use of abseiling techniques for cleaning the glass.

[Above] *Use is made of the wave-form factory roof to control air movements within by positioning extracts for used air at each apex.*

[Left] *A view of the dramatic canopied entrance bridge into the Dyson HQ.*

[Below] *A wave-form roof with steel profiled roof deck provide an industrial aesthetic to working areas of the Dyson HQ in Malmesbury.*

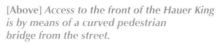

[Above] *Access to the front of the Hauer King is by means of a curved pedestrian bridge from the street.*

[Left] *A long slope of glass forming the roof at the rear of the Hauer King House.*

[Above] *Model of the Croydon Bridge project – a design developed with Future Systems.*

[Right] *The Metropolitan Mast, an asymmetric tower to be built overlooking west London with Future Systems.*

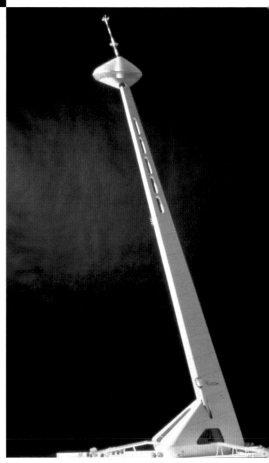

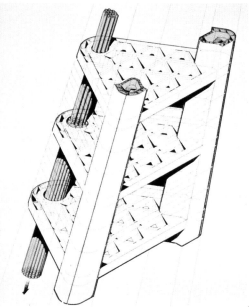

[Left] *Cut-away section through the Metropolitan Mast showing its construction.*

[Below] *Model of the glass-roofed competition entry that Hunt designed with Future Systems for the Stonehenge Visitor Centre.*

The year 1996 brought Hunt a commission for the design of West India Quay Footbridge – a floating bridge. This project and the Hauer King House reunited Tony Hunt with Jan Kaplicky. Kaplicky had worked for Norman Foster early in his career, but had become a founding partner of Future Systems which was established in 1979. Future Systems were omnipresent throughout the High-Tech era, providing striking images for lightweight buildings (often machine-like objects), using imagery derived from motor vehicle, aircraft and spacecraft design concepts. By the time of these two projects, Amanda Levete (who Tony Hunt knew from her time at YRM architects) had joined Kaplicky at Future Systems. Kaplicky died suddenly in his native city of Prague in 2009. The West India Quay Footbridge is a pontoon bridge, floating in the still waters of the dock. With floats resembling those of seaplanes, and perhaps with the images of Bailey bridges in his mind's eye, Hunt was able to respond to the demands of architects' futuristic designs once again.

[Above] *The elegant floating footbridge viewed from the waters of West India Quay.*

[Below] *The crisp lines of the West India Quay floating footbridge seen from above.*

[Opposite] *A photographic simulation of the floating footbridge with the central span lifted to allow access to the dock.*

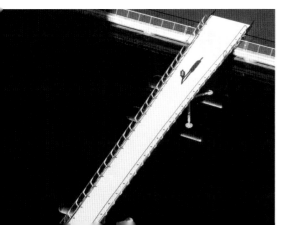

Amongst the magnificent images that Future Systems produced was the Metropolitan Mast – an asymmetric tower to be built overlooking west London in the unlikely setting of Kensington Gardens. There was also a sunken, below-ground Visitor Centre proposed for Stonehenge and a scheme for a new Croydon Bridge. Tony Hunt personally engineered these futuristic structures. As Future Systems projects were schematic, they can only be classed as speculative as far as any commercial benefit might be concerned. With Hunt, Neil Thomas of AHA worked up a fully detailed design for the Mast structure.

Further commissions followed and 2001 saw a commission for the Mount Stuart Visitors' Centre on the Isle of Bute. The architects for this project were Munkenbeck and Marshall. The client was John Crichton, the 7th Marquis of Bute – who is known as Johnny Bute, but who was formerly known as Johnny Dumfries, the 1988 LeMans-winning Jaguar racing driver.

The National Botanic Garden of Wales glasshouse is on a grand scale and an equal to Joseph Paxton's Great Conservatory at Chatsworth House and Decimus Burton's Palm House at Kew Gardens in London.

[Opposite page] The floating footbridge that corresponds to the Docklands street pattern.

With commissions for work at the Eden Project near St Austell, Cornwall, and the National Botanic Garden of Wales in Camarthenshire, Tony Hunt was able to contribute to the great British tradition of glasshouse design and construction, which had design sources back to Joseph Paxton's Great Conservatory at Chatsworth House and Decimus Burton's Palm House at Kew Gardens in London. At the Eden Project – working once again with architects Nicholas Grimshaw and Partners – batch-produced components (albeit of European-origin made by Mero Structures) were assembled to construct the biospheres. These – the largest plant houses in the world – were designed in the image of a group of Buckminster Fuller's geodesic domes. Hunt has claimed that the weight of the structure is less than the weight of the air contained within. This was another triumph of engineering that owed much to the vision of the founder of the Eden Project, Tim Smit – with whom Tony Hunt struck up an immediate and lasting friendship. The main difference between the Eden Project domes and Buckminster Fuller's versions is that Eden has two layers of structure, which makes the structure a space frame. Fuller's domes were designed as a single layer of structure. Buckminster Fuller once famously said: 'nature does not use scaffolding'. The Eden Project had a need for scaffolding, and indeed held the record for the greatest amount of scaffolding for any job in the UK. A conscious decision had been taken by Smit and his advisors to appoint the same design team that had successfully worked together at the Waterloo International Terminal project.

In 2001 amid extensive publicity, the Eden Project with its two huge biodomes, opened as a visitor attraction, boasting the largest greenhouse in the world. Despite being a long-term horticultural development, its immediate popularity could be put down to the impressive scale and bold engineering of its buildings. Each of the spectacular biodomes comprises a series of four interconnected domed structures (geometrically generated from spherical caps) of various sizes. The biggest – the Humid Tropics Biodome is 55 metres high, 100 metres wide and 200 metres long. It covers 15,590 square metres and is designed to maintain a temperature between 18 and 35 degrees Celsius. The air inside is kept moist by abundant water movement, including a waterfall. The smaller Warm Temperate Biodome is 35 metres high, 65 metres wide and 135 metres long, and is designed for temperatures maintained above 10 degrees Celcius in winter and between 15 and 25 degress Celsius in summer months. The domes are constructed from galvanised steel framework, supporting a series of hexagonal panels. As glass would have been too heavy and dangerous for such an application, a special air-filled panelling system was devised. Triple layers of ethylene tetrafluoroethylene (ETFE) foil – transparent to ultra violet light – were stretched across the frames, some over 11 metres across. With an expected life of 30 years, the panel can only be maintained by abseiling along the panel joint lines.

The history of engineering for the Eden Project can be traced back to a number of sources. First of these is the tradition of British glasshouse construction including the Great Conservatory at Chatsworth and the Palm House at Kew Gardens. Secondly, the work of Richard Buckminster-Fuller, who promoted the ideas of geodesic domes in the

[Right] *Publicity image of linked domes, prior to construction.*

[Below] *Dramatic image of the Humid Tropics biodome, kept moist by abundant water movement.*

[Below right] *CGI representation of arrangement of the hexagonal panels.*

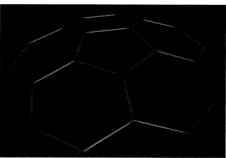

[Above] *Photograph during construction of the biodomes, showing the combined ground works and superstructure into the site of an old china clay pit.*

[Below] *The same view with construction substantially complete.*

[Opposite] *Arrangement of the hexagonal panels of triple layered ETFE foil to give the domed surface of the Eden Project biodomes.*

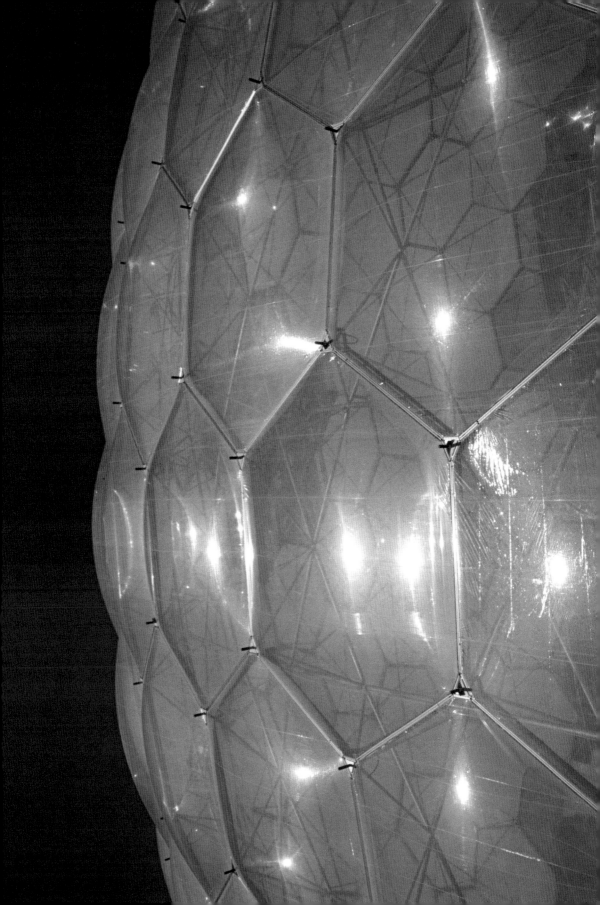

1960s and who was responsible for the USA pavilion at Expo '67 in Montreal, Canada. Thirdly – and perhaps most significantly – from the work of Frei Otto, the German architect who studied the interconnection of domes by inspecting the way that soap bubbles form together. Frei Otto designed the tent-like shapes of the German Pavilion also at Expo '67 in Montreal. It is appropriate that the company with which Otto then worked, Mero Structures, were the steelwork contractors for the Eden Project. Although the biodomes resemble geodesic domes, they owe more to the inspiration of the Victorian engineers in Britain and to the pragmatism of today's German engineering, than they do to the Buckminster-Fuller dream. Both Frei Otto and Buckminster Fuller were lecturing at Washington State University in St Louis in the late 1950s, confirming that they shared a similar vision for lightweight tensile structures. Fuller coined the phrase 'tensegrity' structures to describe a lightweight structure in which tensile cables and rods were used throughout.

Another glass roofed plant house building, opened in 2000, was the National Botanic Garden of Wales. A vaulted arched toroidal roof generated a glass house of crisp geometric form, similar in essence to Ralph Freeman and Ralph Tubbs' Dome

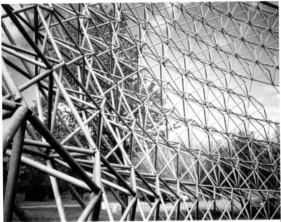

[Left and below] *The authentic Buckminster Fuller geodesic dome from Expo '67, Montreal.*

[Opposite page] *Sketches from Tony Hunt's hand offering several alternative structural grids for the National Botanic Garden of Wales.*

of Discovery built for the Festival of Britain in 1951. This had none of the complex geometry of interconnecting spherical domes required at the Eden Project. A meeting took place in the early stages of design for the National Botanic Garden of Wales. Spencer de Grey – the stalwart partner of Foster's at Foster Associates – met with Alan Jones and Tony Hunt of AHA to consider the possibilities of proving designs for the roof structure. The indications were that such a toroidal shape would be possible with a single layer of structure. Possible sources of difficulty were identified, such as the condition where there was a localised yet heavy snow load. However, Norman Foster

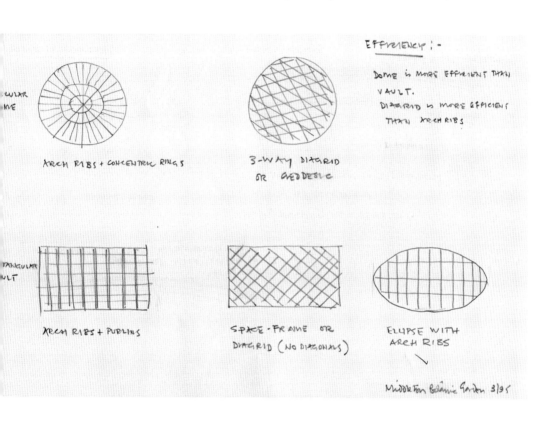

himself was sceptical and suggested that German engineer Jorg Schlaich (b.1934) should be included in the design team. Norman Foster had sought Schlaich's advice on a variety of subjects, and was keen to bring Schlaich into the discussions. In the face of what might have become a threat to Tony Hunt's reputation, with the possible public expression of doubts that Foster may have had about his competence, Hunt threatened to resign immediately. Norman Foster backed away from this confrontation, and the resultant structure – surely one of the purest structural forms of Hunt's entire career – became testament to his command of lightweight structural design, as well as, when necessary, his willingness to be outspoken in defence of his own design ability and reputation.

[Right] *In course of construction, transverse beams are located into ball-joint connectors.*

[Below left] *The vaulted, two-pin arched, toroidal form of the glass roof of the National Botanic Garden of Wales.*

[Below right] *The ball-joint – refinement to the ultimate degree of a structural joint.*

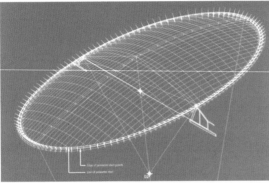

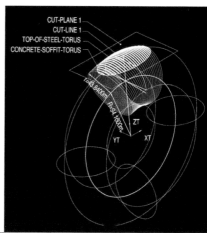

[Above] *The crisp geometry of the National Botanic Garden of Wales roof glinting in the sunlight.*

[Right] *The symmetrical toroid form of the National Botanic Garden of Wales roof is canted to an angle of 7.2 degrees to the horizontal, allowing a 'high side' for tunnel-type access.*

[Above right] *CGI of a toroid (sometimes called an 'anchor ring'), with the cut-off plane required to generate the structural form.*

[Below] *Distant view of the National Botanic Garden of Wales glasshouse.*

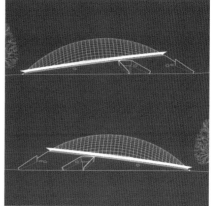

11 (PART I)   HUNT'S REDEFINITION

Hunt felt at home in London in the 1960s. Covent Garden, where his offices were located, was not then the tourist hotspot that it is today. It was more rough and ready, not yet gentrified – there were unused affordable warehouse buildings. Pop culture had its routes nearby and Hunt's passion for life took him to numerous parties and events. He has always enjoyed cinema from his time as a child in London. It was Flash Gordon and Buck Rogers that had entertained the youth of Hunt's generation.

A number of characteristics define Tony Hunt's position within the design community of the early days of AHA. In 1960 the film *Never on a Sunday* portrayed images of free spirits in the Greek islands. Tony had considerable empathy with the sentiments of this film. His manner was one of relaxed authority in his mastery of lightweight structural design. Of the films of the 1960s, *Barbarella* (1968) represented the quest for fun – the same sort of quest as was evident in Mike Webb's Sin Centre and Cedric Price's Fun Palace. Scenes from the film show Jane Fonda crawling along a translucent ribbed tunnel structure, not unlike the external escalator shafts on the Piano and Rogers designed Pompidou Centre. Another film that portrays pop culture in London in the 1960s more accurately is Michaelangelo Antonioni's *Blow-up* (1966) the winner of the *Palm d'Or* in the 1967 Cannes Film Festival. David Hemmings played the part of a fashion photographer resembling a David Bailey character. The film evokes images of a world of studios, clothes and models within the bleakness of urban London. The plot of the film is to do with a crime which may have been captured inadvertently by camera shot. The hectic and panicky story depicts a world of vivid photographic images, in this case enriched by drugs. Pleasure is sought in a whirl of urgency. The film gave insight to what was only a thin veneer of glamorous creativity. Under that surface lay a harsh 'dog eat dog' battle for prestigious work – work which creative people needed in constant supply to establish and enhance their own reputations. AHA's move to Coln Manor in the tranquil surroundings of the Cotswolds, represented a move away from the hectic lifestyle demanded of creative people in London. Fashion, photography and music can respond easily and quickly to whims and gimmickry, but Hunt's opinion was that structural engineering was more serious and needed a stable, conservative and unhurried environment in which to thrive. There was an abundance of vividly displayed architectural images in London of the 1960s, and Hunt's vision for structures in response to these demands was itself a highly charged creative activity. There was a wide-held belief in the London of the 1960s (as there is today) – especially in the architectural profession – that London was the only place in the UK for an architect to establish his or her reputation. The Architectural Association, with its unrivalled history, is in Bedford Square, London. Naturally, it was taken as read that structural engineers should follow suit. It was, therefore, a bold move by Tony Hunt to take his talents away from the creative hot-house of London. He was redefining his own work/life balance. The Cirencester area proved to be a magnet throughout Hunt's life. With his third marriage to Diana Collett (this having started with their meeting in London), it wasn't long before they moved to live in Gloucestershire.

Under Tony Hunt's guidance, his staff had developed a computer system for the calculation of structural performance in the time at Coln Manor between 1976 and 1985. This had been a structures calculator rather than a graphic system to view the objects being designed. Tony Hunt has conceded that it was only possible to design and construct the huge biodomes of the Eden Project and the snaking tapering train shed of Waterloo International because of the advances made in computer modelling techniques. These techniques became available and were used for the Eden Project in conjunction with Mero Structures' experience of structural systems and component manufacture. These telling remarks allow some understanding of the nature of the change that the architectural and engineering profession underwent in the 1990s. The abiding image of Tony Hunt as the structural engineer of the 1960s and the 1970s – the 'enabler' for architects with their vivid imaginations, and his relaxed authority on the subject of lightweight structural analysis, must be redefined by a new image of Tony Hunt – the high profile public face of a multidisciplinary office. By managing this role, he was able to head a design team now capable of resolving the design issues that projects such as the Eden Project might demand. The 'hands on' methods that Hunt had grown up with would be replaced by a more remote overview of engineering and of staff who would be specialists in a more involved, intricate and complicated design process. Hence his adaptation from an individual identity as a structural engineer to the emergence of a 'team player', (which would ultimately allow his retirement to take place) was complete.

Tony Hunt during construction of the Eden Project.

11 (PART II) HUNT'S LASTING INFLUENCE

Due to the unique personal relationships forged at the inception of the High-Tech Movement, Tony Hunt can be said to have presided over the early stages of this movement, and was instrumental in the successes of Team 4, Rogers, Foster, Hopkins and Grimshaw throughout the 1960s and 1970s. It can be argued that Hunt's consistency in his design solutions may have worked against his continuing involvement with the High-Tech architects as they expanded their scope of work nationally and then internationally. Designs of steel lattice beam and square hollow section columns were used in projects for Foster Associates (IBM Pilot Head Office, Cosham 1971), for Piano and Rogers (Universal Oil Products factory 1974), and for Hopkins (Hopkins House 1976). It may be that as these architects strived to identify and promote their own particular brand of architecture, Hunt was caught in a position where his lightweight steel structures became bland and not sufficiently tailored to the needs of each architect. He had fallen victim to the 'possessiveness' that architects felt for their designs. The attitudes of clients and funding organisations, as well as the building regulation authorities, worked against lightweight structures. Exposed and unprotected steel structures were in truth and at best, limited to the occasional single-storey building. Some manoeuvring was possible – such as the use of an upper floor entrance (claiming the lower floor was a basement), or the use of free standing mezzanine floors – and these might mitigate the problem to a limited extent, but the reality was that for the majority of buildings – particularly those in city locations – lightweight structures were very difficult to justify to clients, their funding organisations, and the building regulation authorities.

[Right] Universal Oil Products 1974 building – a post-and-lattice beam single-storey repetitive structure clad with uniformly sized panels.

[Opposite] Column-to-beam connections of the Universal Oil Products factory, with the columns set at 45 degrees rotation to beams for ease of connection.

Hunt's relaxed command of structural analysis, attention to and appreciation of detail and his ability to identify and express key components (and his knowledge of the technology required for their realisation) set him apart from his peers. As such, comparisons with Charles Eames and Jean Prouvé are vital in understanding the context of Hunt's contribution. Knowledge of materials learnt from hands-on prototyping and model making were central to Hunt's design abilities, as they had been to Eames and Prouvé's. Hunt's beautifully-maintained sketchbooks from all stages of his career are testament to his delight in the design process.

There are numerous projects from Tony Hunt's working life that will feature as important buildings for future generations. Naturally, the Grade I listed examples such as Willis Faber Dumas in Ipswich are protected against any risk to their future and will remain as monuments to the co-operation between Hunt and the architects he worked with. However, some of the buildings were only ever considered as temporary, and their future will always depend upon their owners taking care to look after them. The Sainsbury Centre for Visual Arts on the campus of UEA is an example of continuing care needed to the fabric of High-Tech building. Since its completion in 1978, there have been major extensions and refurbishments, during which the panels were removed and replaced with a more technologically-advanced type. The care and attention received by this building is in stark contrast to the nearby UEA campus buildings and walkways designed by Denys Lasdun, which are now showing their age. 27 Broadley Terrace – the offices of Michael Hopkins and Partners built in 1984 (comprising a relocated Patera Building and the larger part built from similar components) – were originally built on a temporary Planning Permission. Again, due to continued care and attention, this complex has lasted more than 25 years. Another example would be the IBM 'temporary' HQ at Cosham, which has been in continuous use since 1971. We can add to this list Richard Horden's Ski Haus, located in the most extreme and hostile environment, which has functioned well for nearly 20 years.

How disconcerting it is therefore, when projects that have received critical acclaim and have been a success in design terms are torn down. We can take this feeling of outrage back to the Festival of Britain where in 1952 – the year after the exhibition – several buildings from leading designers were demolished. We can even take the example of the Sir Herbert Baker designed Bank of England rebuilding in 1923–1933, which replaced the original building (c.1800) of Sir John Soane. Certainly we should include the Brynmawr Rubber Works in South Wales – a tour de force of reinforced concrete design by Ove Arup and his structural analyst Ronald Jenkins, which despite its Grade II* listing, was demolished in 2001. Despite its great critical acclaim, Reliance Controls – built in 1967 – was demolished in the 1990s to make way for renewed property development. When first built, its design by Foster Associates and AHA heralded a newness in industrial architecture. Using standard industrial components in a new way (by expressing structure and celebrating its fine proportions), Foster and Hunt set the pattern for future industrial building. The Reliance Controls building did not receive the care and attention which would have been required to maintain its pedigree. Indeed, at one stage the carefully proportioned bays were 'vandalised' by the insertion of incongruous standard domestic off-the-peg timber casement windows, to suit some new internal arrangements. This lack of acceptance and disregard for design is akin to that which sealed the fate of the Bean Hill housing development in Milton Keynes. The difficulty of gaining approval for industrially harsh materials in the cosiness of a home environment can equally be applied to small workshop applications. Foster and Hunt's early work for the Fred Olsen Line – both the Passenger terminal and the Olsen Centre – have also been demolished. At one point, Tony Hunt referred to his wish to form a small exclusive society of architects and engineers who have outlived some of their best works.

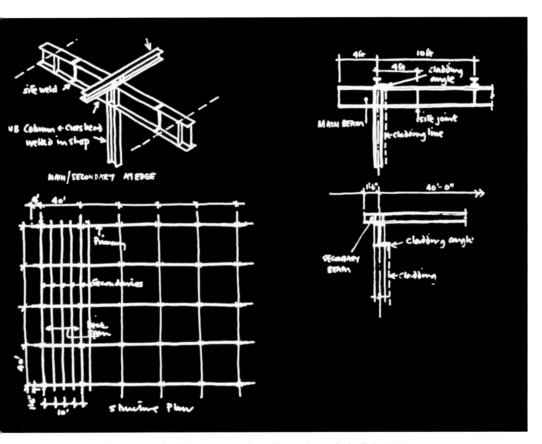

Tony Hunt's sketch of the structure for Reliance Controls, built in 1967.

Further lasting influence exists in a number of Hunt sketches, some of which have recently been exhibited at the Chelsea Art Club. The club have traditionally produced a yearbook, to include sketches by members and other contributors. In 2009 the club mounted its first exhibition of architectural sketches, in which Tony Hunt's work was included.

[Right] *Standard industrial components used in a new way provide a finely proportioned and elegant design solution (exemplified in Reliance Controls) to industrial building.*

[Opposite] Foster Associates' cross-sectional centre-perspective of the Fred Olsen Lines building – a graphic technique used widely at the time.

There are a number of phrases that have become attributed to Tony Hunt over the years, and whether or not he was originally responsible for them, they do give an insight into his design philosophy.

'Concept before calculation.' This would be clearly demonstrated in almost all of Hunt's designs, when he says that he would not want an architect to present him with a near complete set of drawings for him to engineer. Rather, he would (in discussion with the architect) consider a range of structural principles and of preferred materials, establish some sort of structural form possibly with an outline of a structural grid, and only then would he start to evaluate the shapes and sizes of any structural members. Calculation would come when the structural principles were established and the calculation could then be focussed on refinement, the finding of the most efficient structural members and the most appropriate connections between them.

'No preconceived ideas.' Here, Hunt would be saying that he was willing to work with architects to fulfil their imagined schemes, whatever the architectural style. Throughout his six decades of practice, he has seen many styles and architectural movements come and go. He has worked with steel, concrete, wood, aluminium and GRP structures. He has worked with foundries, rolling mills, smelters, extruders, pattern-makers, mould-makers, formwork-makers, machine shops and concrete-pourers. Understanding the technology behind the manufacture and processing of materials enabled Tony Hunt to approach each project with an open mind as to what the most appropriate structure might be. He has been able to work with public sector or private clients and with architects in most continents. Hunt has been able to work with contractors, co-partnerships and even with the most informal of contractual arrangements such as the Crestone Dome, which is essentially a quasi-religious building constructed using volunteer seasonal labour over a period of three summers. David Tasker – a former employee of Hunt's from AHA's early London days – supervised the construction of Crestone.

'Easy communication.' Hunt's relaxed style, his experience and authority set the tone for any bilateral design dialogue between architects and his engineering. He is able to convey ideas by sketching out structural concepts with an uncomplicated fluency that sometimes belied the complexity involved. When asked the question, 'Have you ever designed and built a house for yourself?', Tony Hunt answers no, but if he ever was to build for himself it would be a component-based design similar to his view of furniture design. This is the aptitude that Hunt has always managed to preserve: that he will respond to any architect in the design language and vocabulary that they bring to the table. It may be masonry, timber, steel or aluminium, it may be wide clear spans or grid-like short span structures, but whatever the brief, the design will receive a sympathetic response to make the most effective structural statement. Hunt's first instinct will be to listen and only when he has heard the force of prevailing argument from his client will he offer his design solutions.

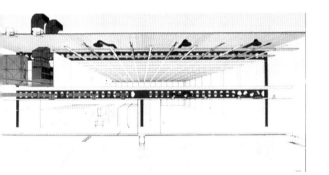

In the matter of glass buildings, there must have been a note of regret that Hunt felt over the abandonment in 1994 of the Centre du Conferences Internationales de Paris (CCIP). This was a design competition won against fierce opposition by architect Francis Soler in 1990 – a part of President Mitterand's grand vision for Paris. The successful design was to have been built in the shadow of the Eiffel Tower and comprised of three rectangular glass halls each of 50 metres clear span, 150 metres long and 35 metres high. Comparisons to Paxton's Great Exhibition building of 1851 would have been inevitable – a new Crystal Palace.

When in the mid 1980s the role of the engineer changed to being project-based instead of providing industrially-designed solutions, the personal relationships made with architects and with clients set Hunt in a strong position to adapt to the change. He was able to concentrate on a series of prestigious projects. The size and complexity of projects meant that small jobs – typically a personal project such as a house or studio – could no longer sustain either architectural or engineering offices. The design of houses or studios became little more than a hobby.

In the six decades of Hunt's career, he has seen many architects come and go, and several architectural movements come into fashion. With direct links to the British Modern Movement, through the High-Tech and the Post-modern Movements, and to today's super budget landmark projects, Hunt's support of architects has always been welcomed. The lessons of collaboration learnt at Samuely's set him in good stead throughout his professional life. His passion for industrial design set him apart from his peers.

In the run-up to the millennium, Tony Hunt had the opportunity to become one of the judges in a design competition for a tower to be a part of the Clydeside development – the Glasgow Science Centre. This gave Hunt the chance to be involved with the processes of design from the patrons' side. Up until this point he had assisted with entries for such competitions on numerous occasions. Unusually, Hunt was able to show his preference for lightweight innovative sophisticated steel structures, not as an engineer promoting such designs to architects, but as an impartial arbiter of the designs placed before him. The selection of the winning entry might dictate that the designs would be from an architect that Hunt had not necessarily worked with in the past, and the structural engineering would be by a rival engineer with whom Hunt would normally be competing. The Glasgow Tower – completed in 2001 – is a prize-winning design by architect Richard Horden, the result of a design competition. Tony Hunt with Norman Foster and others sat on a jury to decide which design should be selected. Horden's design for a 'winged tower' 127 metres high included a viewing platform, restaurant and exhibition space. With aeronautical engineer Peter Heppel, Horden proposed the use of a vertical wing that could rotate about its vertical axis to suit the wind direction. It would be capable of rotating throughout 360 degrees (the first to be able to do so), in order to reduce the wind loading on the structure. Buro Happold were appointed as structural engineer for the project.

Considered one of the High-Tech architects, Richard Horden is an English architect who derived inspiration from the elegance of yacht design and who adopted the

The Richard Horden designed tower, a vertically pivoted wing which responds to the prevailing wind direction. (Image courtesy of Brian Negus)

principles of his contemporaries in his use of batch-produced industrially-prefabricated components. In 1984 Horden had built a house for his sister at Woodgreen, Hampshire. The structure for this house was based upon a square grid of columns and beams adapted from aluminium extrusions intended for yacht masts. He was able to transfer the technology from the yacht building industry. In the case of the Horden House, the structural analysis to transfer the yacht technology for use in buildings was not down to Tony Hunt, but his erstwhile mentor at Samuely's – Frank Newby. Tony Hunt had sailed with Richard Horden on at least one occasion (Hunt remembers sailing a catamaran; they both shared a passion for this pursuit). Comparisons can be made between Richard Horden's house and the unbuilt Norman Foster prototype house. Structural connections were compromised by jointing the oval cross-sectioned mast components used in the Horden example whereas, by using steel inserts, Hunt was able to devise a much more elegant solution for Foster.

Hunt believes strongly in sharing his experience and knowledge. He lectures regularly in the UK, Europe, USA and Canada. Tony Hunt is a Fellow of the Institution of Structural Engineers and was a Gold Medalist in 1995. He is also an Honorary Fellow of the Royal Institute of British Architects. He holds Honorary Doctorates from the Universities of Sheffield (1999) and Leeds (2003).

He was a visiting professor at the Universities of Pennsylvania, Sheffield, the Chinese University of Hong Kong, IST Lisbon and Birmingham. Hunt is currently a visiting professor at KAA (Royal Danish Academy of Fine Arts, School of Architecture, Copenhagen). He has several published works including *Tony Hunt's Structures Notebook* (the Architectural Press, 1997) and a second edition in 2003 and an English/Chinese edition published in 2004) and *Tony Hunt's Sketchbook* (the Architectural Press, Volume 1 in 1999 and Volume 2 in 2003).

Many practices have been spawned by staff that gained their experience whilst working for AHA. Carter Clack set up independently after AHA started to expand and the move to Coln Manor was completed. Adams Kerr and Taylor (AKT), Stephen Morley (Modus), Matthew Wells (Technica), Bob Barton (Barton Associates), Mark Whitby (who established Whitbybird after a short time with Buro Happold establishing their Bath office), Brian Forster (who went to Arups), and Neil Thomas (Atelier 1) all gained valuable experience with Tony Hunt. Furthermore, Wolf Mangelsdorf left for Buro Happold, and Les Postawa left the AHA London office to work for Thornton Tomasetti. Hunt's style of teaching is one of informality, of demonstration, and of dialogue with clients, architects or other engineering students. He is not in the least dictatorial, but will always draw out opinions other than his own in order to reach a conclusion. He will not be hurried at the formative stages. His design will always have come about by narrowing down many different options for a given set of circumstances. Not until he has investigated all possibilities will he commit to a given solution.

In his lectures – which are always a visual presentation accompanied by his own narrative – he will explain how any particular design was arrived at. Studio-centred master-classes would be his preferred setting for teaching. This has been his most recent experience at KAA, where he has met up once again with Olga Popovic Larson – a Lecturer in architecture that Tony met first at Sheffield University School of Architecture. Olga Popovic Larson has specialised in researching timber-frame structures and geometrically progressive structural grids. She has written a book: *Reciprocal Frame Architecture*, which contains a foreword written by Tony Hunt.

Tony Hunt has always taken an interest in the history of engineering, again following the lead of Frank Newby, Julia Elton and others. He met Sonia Rolt – widow of the acclaimed writer L.T.C. (Tom) Rolt – on the occasions when he attended meetings of the History Study Group of the Institution of Structural Engineers. James Sutherland was the first driving-force behind this group, with Frank Newby taking over in 1993.

Tony Hunt retired in 2002 assigning his professional interests to SKM Anthony Hunt's. Sinclair Knight Merz (SKM) are an international multidisciplinary engineering company originating from Australia. Tony Hunt remains a consultant to them. Hunt recalls that one of the last projects that he worked on with SKM was the Camden Roundhouse Development with architect John McAslan. He experienced the delight of surveying the original brickwork structure, which had been designed to carry the turntables of the 1847 railway steam engine shed. He recounts how the head of the Camden Roundhouse Development – Sir Torquil Norman (of the Bluebird toys firm) – was able

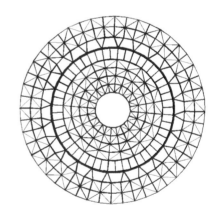

to find the funding and steer the development through in the face of opposition from other development proposals. When it was first built, the Camden Roundhouse structure belonged to the London and Birmingham Railway, which operated out of Euston Station. I. K. Brunel was engineer for the GWR, which did use Euston Station before Paddington was completed, despite the differing gauge of their tracks. It is not thought that the roundhouse is a Brunel structure, despite having all the hallmarks of such pedigree including an exquisite brick-arched undercroft.

With John McAslan, SKM engineered a project for Kelvin Bridge in Glasgow. With Wolf Mangelsdorf, Tony Hunt took a close interest in the design of a curved footbridge made from steel and glass. The bridge deck itself was to be made of glass. As with many futuristic designs, this bold scheme was unsuccessful, and was never built. There are three other projects from the last period of Hunt's involvement at SKM that he remains interested in. These are Linear City in Kuala Lumpa; a bridge constructed by local labour in China; and a glass bridge that was designed for the docklands area of London, but which now may be re-sited.

Working with architect Edward NG Yan-yung , professor at the Chinese university of Hong Kong , Tony Hunt engineered a modest but effective bridge re-building over

[Above] *John McAslan Architect's imagery for the Camden Roundhouse used in development promotional material*

[Left] *The fly gallery of Camden Roundhouse showing the new steel structure introduced to support the existing roof.*

[Below] *An isometric 3-D CGI demonstrating the complex geometry of the steelwork to the roof.*

[Opposite] *An intricately resolved steelwork diagram for the structure to be introduced into the existing roof of the Roundhouse.*

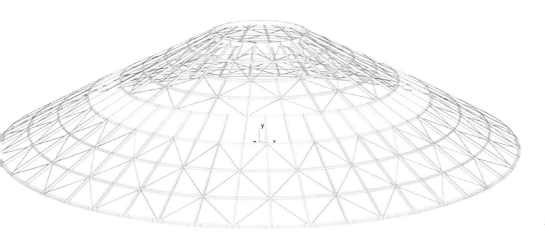

[Right] *Rhythmical series of spinal 'Y'-shaped columns, of gently changing palette.*

[Opposite] *The end elevation expresses the wave form of the roof and the 'Y' columns which form the spine of the terminal.*

the River Po, a tributary of the Yellow River, where it runs through the village of Wuzhi in the northern western province of Gansu. Known as the Wuzhi Qiao bridge and completed in 2005, it is 90m long and divided into 20 sections. Each section spanning between crated stone piers is made from prefabricated steel and bamboo bridge beams; it was made by locally sourced materials worked by simple tools.

A further project that Tony Hunt worked on late in his career, but which was completed after his retirement was Madrid's Barajas Airport. This was a major project designed by Richard Rogers Partnership. The design won the Stirling Prize in 2006. The design is notable for its central spine of 'Y' shaped columns which support an 'M' shaped wave-form roof (in cross section the 'M' is rounded and flattened), with the forks of the 'Y' supporting the wave form symmetrically. At 1.2 km long and with the columns subtly painted in different colours depending upon sectors or zones, the impression is almost hypnotic: as close to Tubbs' 'mathematical poetry' as any of Hunt's structures.

[Right] *Cannopied walkway throughout the length of the airport terminal.*

[Below left] *Presentation image of Glass Bridge.*

[Below right] *Representation of Gigaworld – a complete linear city raised above a river and running for several miles in length.*

[Above] *Hélène off-duty from the kitchen enjoying seafood cooked for her.*

[Left] *Tony in celebratory mood at the Seymour-hosted party.*

After retiring, Hunt moved briefly to a farm in Gloucestershire with Diana, but since 2007 he has lived in the familiar surroundings of the Cotswolds after his third marriage (of 22 years) ended. He now lives with retired restaurateur Hélène Moore.

In 2007 Ed Seymour, of Taylor Seymour Architects in Malmesbury, hosted a grand party to celebrate Tony's 75th birthday. Seymour took the opportunity to renovate a courtyard area adjacent to his house in time for the event. Tony explained that his studio property in Box does not lend itself to 'a decent party' as the accommodation is split over three floors.

12 CHRONOLOGY AND SELECTED WORKS

1951	Anthony Hunt joins F.J. Samuely & Partners
1959	Cantu Furniture Design competition, honourable mention
1959	Hunt joins Terence Conran's firm
1960–62	Hunt joins Hancock Associates
1960	Synagogue roof, Carmel College, Oxfordshire (architects: Hancock Associates)
1962	Anthony Hunt Associates formed
1962	Nereid Monument, British Museum, London (architects: Russell Hodgson & Leigh)
1962	Leicester University Library, Structair system, Leicester (architects: Castle and Park)
1964	Creak Vean house, Feock, near Falmouth, Cornwall (architects: Team 4)
1966	Murray Mews houses, London (architects: Team 4)
1967	Reliance Controls Factory, Swindon, Wiltshire (architects: Foster Associates)
	Financial Times Award for Industrial Architecture, 1967
1967	Newport High School, Gwent, South Wales (architects: Evans and Shalev)
	Competition Submission-Prize Winner, 1st Prize 1967
1968–71	Zip-up house project (architects: Richard and Su Rogers)
1968	Rogers house, Wimbledon, London (architects: Richard and Su Rogers)
1968	Spender house, Utling, Essex (architects: Richard and Su Rogers)
1970	Olsen Lines Operation-Amenity Centre, Millwall (architects: Foster Associates)
	Financial Times Award for Industrial Architecture, 1970
1971	Factory for Computer Technology, Hemel Hempstead, Hertfordshire (architects: Foster Associates)
	Financial Times Award for Industrial Architecture, 1971
1971	Fred Olsen shipping line passenger terminal and HQ, Millwall (architects: Foster Associates)
1971	New Parliamentary Extension, Westminster, London (architects: Spence and Webster
	Competition Submission-Prize Winner, 1st Prize 1971
1971	Aram Standardised Hospital Module, prototype, USA (architects: Piano and Rogers)
1971 (*circa*)	Field House, Crocknorth, Surrey (architects: Georgie Wolton)
1971	IBM Pilot Head Office, Cosham, Hampshire (architects: Foster Associates)
	Structural Steel Design Award, 1972
	RIBA Award-Winning Building, 1972
1972	Design Research Unit (DRU), Aybrook Street, London (architects: Richard and Su Rogers)
1972	Prototype Advance Factory Unit, Milton Keynes (architects: Milton Keynes Development Corporation)
	Structural Steel Design Award, 1972

1973	Modern Art Glass Building (architects: Foster Associates)
1973	Newport High School, Gwent, South Wales (architects: Evans and Shalev)
1973 (circa)	Pill Wood House, Feock, near Falmouth, Cornwall (architects: Colquhoun and Miller)
1973	Sapa factory, Tibshelf, Derbyshire (architects: Foster Associates)
1973	Sobell Pavilions, London Zoo, London (architects: John Toovey)
	Structural Steel Design Award, 1974
	Civic Trust Award-Winning Building, 1973
1974	MAG warehouse and showrooms, Thamesmead (architects: Foster Associates)
	Financial Times Award for Industrial Architecture, 1974
1974	Offices, Norwood, London (architects: Crowley Moore-Ede)
	Civic Trust Award-Winning Building, 1974
1974	Universal Oil Products, UOP Fragrances, Tadworth, Surrey (architects: Piano and Rogers)
	Structural Steel Design Award, 1975
1975	Willis Faber Dumas building, Ipswich, Suffolk (architects: Foster Associates)
	RIBA Award-Winning Building, 1977
1975	Leicester University Library, Leicestershire (architects: Castle, Park Dean and Hook)
	RIBA Award-Winning Building, 1975
1975	Bean Hill housing, Milton Keynes (architects: Foster Associates)
1976	Highgate Wood School Sports Hall, London (architects: Chapman Lisle)
	Structural Steel Design Award, 1976
1975–76	Hopkins house, Hampstead, London (architects: Michael Hopkins and Partners)
	RIBA Award-Winning Building, 1977
	Civic Trust Award-Winning Building, 1979
1976	van den Bossche House, Fluy, France (architects: Ian Ritchie)
1976	English Industrial Estates Corporation, Factory Design (architects: Nixon and Kaplicky)
	Competition Submission-Prize Winner, 1st Prize, 1976
1977	Palmeston Special School, Liverpool (architects: Foster Associates)
	RIBA Award-Winning Building, 1977
1977	Wedgwood House, Colchester, Suffolk (architects: Aldington Craig and Collinge)
	RIBA Award-Winning Building, 1978
1978	Alexandra Road housing, Camden, London (architects: Neave Brown, Camden Architects Department)
1978	Sainsbury Centre for the Visual Arts, phase 1, Norwich, Norfolk (architects: Foster Associates)
	Finniston Award, 1978
1978	Prototype House for Norman Foster, Hampstead, London (architects: Foster Associates)
1979–81	Racking Plant, Greene King Brewery, Bury St Edmunds, Suffolk (architects: Michael Hopkins and Partners)
	Financial Times Award for Industrial Architecture, 1980
1979	SSSALU Prototype Aluminium Building System (architects: Michael Hopkins and Partners)
1980	Eagle Rock House, near Uckfield, East Sussex (architects: Ian Ritchie)
1980	Distribution Centre & Installation Support Centre for IBM, Greenford, West London, (architects: Foster Associates)

	Structural Steel Design Award, 1980

1980   Newcastle-upon-Tyne Quayside Rehabilitation (designers: Bentley Campbell
        Hunt Murrian)
        Competition Submission-Prize Winner, 4th Prize 1980
1980   Passive Solar Competition (EEC) with Energy Design Group
        Competition Submission-Prize Winner, 2nd Prize, 1980 (no 1st Prize awarded)
1980   Westminster Pier, London (architects: Castle Park Hook and Partners)
        Competition Submission-Prize Winner,
1980   New Bradwell, Milton Keynes (architects: Phippen Randall and Parkes)
        Civic Trust Award-Winning Building, 1980
1980   St Nicholas Parish Church, Guildford, Surrey (architects: Nye Saunders and
        Partners)
        Civic Trust Award-Winning Building, 1980
1980–82 Patera Building, Stoke-on-Trent, Staffordshire (architects: Michael Hopkins and
        Partners)
        Structural Steel Design Award, Commendation 1983
1981   Timber dome, Crestone, Colorado, USA (architects: Keith Critchlow)
1981   Kennard Street Health Centre, Victoria Dock, London (architects: Aldington and
        Craig)
        Civic Trust Award-Winning Building, 1981
1981   Architects' Journal, Holiday Homes Competition (architects: Fielden Clegg
        Design)
        Competition Submission-Prize Winner, 1st Prize, 1981
1981   Library, Royal Military College of Science (RCMS), Shrivenham, Oxfordshire
        (architects: Evans and Shalev)
        Competition Submission-Prize Winner, 1st Prize 1981
1982   INMOS microprocessor factory, Newport, Gwent (architects: Richard Rogers
        and Partners)
        Institution of Structural Engineers Special Award 1983
1982–85 Schlumberger Research Centre, Cambridge (architects: Michael Hopkins and
        Partners)
1983   Halley Bay, British Antarctic Survey, Antarctica (architects: Jamieson Associates)
1984   Offices, 27, Broadley Terrace, London NW1 (architects: Michael Hopkins and
        Partners)
1986   Amada Machine Tools factory, Kidderminster, Worcestershire (architects:
        Glazzard Architects Co-operative)
1989   Heathrow Terminal 5 competition entry (architects: YRM Architects and
        Planners)
1989   YRM Anthony Hunt Associates formed
1991   Don Valley Stadium, Sheffield (architects: Design and Build Services, Sheffield
        City Council)
        Institution of Structural Engineers Special Award 1991
        Structural Steel Design Award 1991
1991   Sackler Galleries, Royal Academy of Arts, London (architects: Foster and
        Partners)
        Structural Steel Design Award 1992
1991–93 Waterloo Station, International Terminal, London (architects: Nicholas
        Grimshaw and Partners)
        Institution of Structural Engineers Special Award 1994
        Structural Steel Design Award 1994

1993	International Conference Centre (CCIP), Paris (not built) (architects: Francis Soler)
1994	Hauer King House, Canonbury, London (architects: Future Systems)
1994	Metropolitan Mast, Kensington Gardens, London (architects: Future Systems)
1996	West India Quay Footbridge, Docklands, London (architects: Future Systems)
	British Construction Industry Award 1997
1993–97	Galpharm Stadium (Alfred McAlpine Stadium), Huddersfield, Yorkshire (architects: Lobb Partnership)
	Structural Steel Design Award, Commendation 1995
	RIBA Building of the year 1995
1997	YRM Anthony Hunt Associates disbands, AHA reformed
1997	Stuttgart 21 Hauptbahnhoff competition (architects: Worner and Partner)
1998	Dyson Factory, Malmsbury, Wiltshire (architects: Wilkinson Eyre)
	Civic trust Award, Winning Building 1999
1998	Museum of Scotland, Edinburgh (architects: Benson and Forsyth)
2000	National Botanic Garden of Wales, Camarthenshire (architects: Foster and Partners)
	Structural steel Design Award 2000
2001	Mount Stuart Visitors' Centre, Isle of Bute, Scotland (architects: Munkenbeck and Marshall)
1998-2001	The Eden Project, St Austell, Cornwall (architects: Nicholas Grimshaw and Partners)
	Structural Steel Design Award, Commendation 2001
	British Construction Industry Award, Major Project 2001
	Institution of Structural Engineers Special Award 2002
2001–2002	Anthony Hunt surveys the Camden Round House (architects: John McAslan and Partners)
	Structural steel Design Awards - Finalist 2008
2002	Anthony Hunt retires
2005	China Bridge, Wuzhi Qiao, River Po, China (architects: Edward NG Yan-Yung, professor at the Chinese University of Hong Kong [CUHK])
2006	Barajas Airport, Madrid, Spain (architects: Richard Rogers Partnership)
	British Construction Industry Award - International 2006
	Structural Awards - Commercial and Retail Structures 2006
	Barajas Airport was awarded the 11th RIBA Stirling Prize in 2006

BIBLIOGRAPHY

Banham, Reyner. *Design by Choice*. Academy Editions, 1981.

Banham, Reyner. *Megastructure: Urban Futures of the Recent Past*. Thames and Hudson, 1976.

Bender, Richard (with Foreword by Ezra Ehrenkrantz). *A Crack in the Rearview Mirror: a View of Industrialized Buildings*. Van Nostrand, New York, 1973.

Best, Alistair. 'Two Problems Solved'. *Design Journal,* 1971.

Chermayeff, S & Alexander, C. *Community and Privacy*. Pelican Books, 1966.

Critchlow, Keith. *Order in Space, a Design Source Book*. Thames and Hudson, 1969.

Crosby, Theo. *The Necessary Monument*. Studio Vista, 1970.

Davey, Peter. *Waterloo International Terminal*. The Architectural Review, September 1993.

Davies, Colin. *High Tech Architecture*. Thames and Hudson, 1988.

Davies, Colin. *Hopkins, The Work of Michael Hopkins and Partners*, with essays by Patrick Hodgkinson and Kenneth Frampton. Phaidon, 1993.

Donati, Cristina. *Michael Hopkins*. Milan: Skira Editore S.p.A, 2006.

Drexler, Arthur. *Transformations in Modern Architecture*. Secker and Warburg, 1979.

Earles, William D. *The Harvard Five in New Canaan*. W.W. Norton and Company, 2006.

Edwards, Brian. *Stadium with Swagger*. A.J. Building Study, October 27 1994.

Financial Times Industrial Architecture Award booklet, London: Financial Times Ltd, 1980.

Frampton, Kenneth. *Modern Architecture, a Critical History*. Thames and Hudson, reprint 1982.

Glancey, Jonathan. *New British Architecture*, Thames and Hudson, 1989.

Hunt, Anthony. *Tony Hunt's Sketchbook*. Architectural Press, 1999.

Hunt, Anthony. *Tony Hunt's Structures Notebook*. Architectural Press, 1997.

Jackson, Neil. *Craig Ellwood*. Laurence King Publishing, 2002.

Jackson, Neil. *The Modern Steel House*. E & F.N. Spon, 1996.

Jean Prouvé Constucteur. Museum Baymans-van Beuningen, 1981.

Jencks, Charles. *The Language of Post-modern Architecture*. Academy Editions, 1977.

Jencks, Charles. *Modern Movements in Architecture*. Pelican Original, 1973.

Koenig, Gloria. *Charles and Ray Eames*. Taschen, 2005.

Kron, Joan and Slesin, Suzanne. *High-Tech, The Industrial Style Source Book for the Home*. Allen Lane, Penguin Books, 1979.

Lampugnani, Magnago. *Visionary Architecture of the 20th Century*. Thames and Hudson, 1982.

MacDonald, Angus. *Engineer's Contribution to Contemporary Architecture: Anthony Hunt*. Thomas Telford Publishing, 2000.

Makowski, Z.S. *Steel Space Structures*. Joseph, 1965.

Maré, Eric de and Richards, J.M. *Functional Tradition in Early Industrial Building*. Architectural Press, 1958.

Michael Hopkins: SSSALU (Short Span Structures in Aluminium) Prototype. The Architectural Review, December 1980.

Pawley, Martin. *Theory and Design in the Second Machine Age*. Basil Blackwell, 1990.

Pawley, Martin. *Buckminster Fuller, (Trefoil Design Heroes)*. Trefoil Publications Limited, 1990.

Pearman, Hugh and Whalley, Andrew. *The Architecture of Eden*, Eden Project Books, in association with Grimshaw, Transworld Publishers, 2003.

Peters, Nils. *Jean Prouvé*, Taschen, 2006.

Powell, Kenneth. *New Architecture in Britain.* Merrell Publishers Ltd, 2003.

Powers, Alan. *Modern: The Modern Movement in Britain.* Merrell Publishers Ltd, 2005.

Price, Cedric. *The Square Book*. Wiley-Academy, 2003, copyright Cedric Price 1984.

Richards, J.M. *An Introduction to Modern Architecture*, Pelican Books, reprint 1960.

Russell, Barry. *Building Systems Industrialization and Architecture.* John Wiley & Sons, 1981.

Serraino, Pierluigi. *Eero Saarinen.* Taschen, 2005.

Sheard, Ron. *The Stadium, Architecture for the New Global Culture*. Periplus Editions, 2005.

Sinclair, Cameron and Stohr, Kate (Edited by Architecture for Humanity). *Design Like You Give a Damn: Architectural Responses to Humanitarian Crises*. Thames and Hudson, 2006.

Smith, Elizabeth A.T. *Case Study Houses*. Taschen, 2007.

Spiller, Neil. *Visionary Architecture, Blueprints of the Modern Imagination*. Thames and Hudson, 2005.

Sutherland, R.J.M, Humm, Dawn and Chrimes, Mike. *Historic Concrete: The Background to Appraisal.* London: Thomas Telford Publishing, 2001.

Torroja, Eduardo. *Philosophy of Structures*. University of California Press, 1967.

Treiber, Daniel. *Norman Foster*. E & F.N. Spon, 1995.

Wachsmann, Konrad. *Turning Point of Building, Structure and Design*, Van Nost.Reinhold,U.S, 1961.

Warburton, Nigel. *Ernö Goldfinger: The Life of an Architect*. Routledge, 2004.

Weston, Richard. *The House in the Twentieth Century*. Laurence King Publishing, 2002.

Whitby, Mark. *The Patera Building System. Building with Steel*, Volume 9, No3, June 1983.

Winter, John. *The Patera Building. Architects Journal*, September 1 1982.

Winter, John. *Hopkins House. The Architectural Review*, Volume CLXII, No 970, December 1977.

Winter, John. *Racking Plant, Greene King Brewery, Bury St Edmunds. The Architectural Review*, No169, March 1981.

KEY STAFF (ACKNOWLEDGED BY TONY HUNT AT THE VARIOUS STAGES OF HIS CAREER):

Early days of Anthony Hunt, Consulting Engineer: Leslie Stebbings, Frazer Glegg, Brian Forster, John Austin. Laurie Fogg, David Hemmings, David Tasker, Richard Clack, John Carter.

Later, The first partnership, Anthony Hunt Associates: Tony Hunt, David Hemmings, John Austin, Brian Humphrey (accountant). Associates at that time: Brian Forster, Merlin Saunders, Laurie Fogg, Mark Whitby, Alan Jones, Allan Bernau, Alan Stone, Anthony Wootley.

Joining with YRM as YRM Anthony Hunt Associates: Tony Hunt (Plc Director), David Hemmings, John Austin, Bjorn Watson, Merlin Saunders, Alan Jones, Allan Bernau, Martin Cranidge. Associates: Stephen Morley, Neil Thomas, Bob Barton, Les Postawa, Matthew Wells, David Hamilton, Nick Green (London & Paris). Others: Barry Dooley (Paris) Hanif Kara, Declan Collier, Peter Shepherd, Albert Williamson-Taylor, Robin Adams, Mike Purvis, Jerry King.

The new Anthony Hunt Associates – post YRM. Directors: Tony Hunt, David Hemmings, Alan Jones, Allan Bernau, Les Postawa, Bjorn Watson. Associates: David Hamilton, Martin Jones, Mike Purvis, Gerry Trotter. Key Secretaries, PAs and Admin Staff: Chris Herring, Liz Baker, Gemma Fox (nee Wickham), Dot Warren, Joan Freeman, Lucy Morgan.

INDEX